THE SYNTHETIC '70s

Fabric of the Decade

Constance Korosec &
Leslie Piña

Schiffer Publishing Ltd®
4880 Lower Valley Road Atglen, PA 19310 USA

Library of Congress Catalog Card Number: 98-87850

Designed by Leslie Piña
Layout by Bonnie M. Hensley
Type set in Impress BT/Korinna BT

ISBN: 0-7643-0717-7
Printed in China

Published by Schiffer Publishing Ltd.
4880 Lower Valley Road
Atglen, PA 19310
Phone: (610) 593-1777; Fax: (610) 593-2002
E-mail: Schifferbk@aol.com
Please visit our web site catalog at **www.schifferbooks.com**

This book may be purchased from the publisher.
Please include $3.95 for shipping.
Please try your bookstore first.
We are interested in hearing from authors
with book ideas on related subjects.
You may write for a free catalog.

In Europe, Schiffer books are distributed by
Bushwood Books
6 Marksbury Rd.
Kew Gardens
Surrey TW9 4JF England
Phone: 44 (0)181 392-8585; Fax: 44 (0)181 392-9876
E-mail: Bushwd@aol.com

Table of Contents

Acknowledgments

We would like to express our thanks to the anonymous sources and to the following who so generously provided fabrics for us to photograph for this book: American Fabrics, Lois Epstein, Shirley Friedland, Lillian Horvat, Knoll Textiles, May Brown Korosec, Irene Sladewski, and the Ursuline College Resource Library. Thanks again to Ramón for his indispensable assistance with the photography; to the Ursuline College library staff; to Peter Schiffer, Jennifer Lindbeck, and the gang at Schiffer publishing.

EVOLUTION OF MAN-MADE FIBERS IN THE UNITED STATES

1910
RAYON - Although technically a by-product of cotton, Rayon is also considered to be the first man-made fiber. It was produced 1910 at a plant at Marcus Hook, Pennsylvania, by the American Viscose Company, which had been formed by Courtaulds of England. Made from cellulose, the fibrous substance is found in all forms of plant life. Initially, the cellulose came from cotton linters; however, today most cellulose-derived fibers are made from purified wood pulp.

1924
ACETATE - Acetate fiber in the United States was commercially produced by the Celanese Company in 1924. It is made from cellulose, which when treated with acetic acid and acetic anhydride becomes acetate. Since acetate does not accept the dyes ordinarily used for cotton and rayon, it was necessary to develop special dyes.

1930
RUBBER - Production began in 1930 by the U. S. Rubber Company in the United States. Rubber was replaced by spandex because spandex had better properties than rubber.

1936
GLASS - The Owens-Illinois Glass Company and Corning Glass Works began research and experimentation with glass filaments in the 1930s. The first commercial production of Fiberglas began in 1936. The raw materials used to make ordinary glass are refined and shaped into glass marbles, which are remelted and formed into glass filaments. The resulting fibers and yarns can be woven into strong, durable fabrics that resist heat, flame, and many chemicals. Its uses include home furnishing products such as curtains, draperies, and upholstery, as well as in fire proof clothing.

1938
NYLON - In 1938, shortly before World War II, a new material, nylon, was introduced to the American public. Research was done in the DuPont laboratories. First used in hosiery, the new "miracle" fiber was also used in carpeting and automobile upholstery by the end of the 1940s. Its strength, elasticity, washability, and abrasion resistance enabled its uses to include apparel, home furnishings, and industrial products.

1939
VINYON - The American Viscose Company developed the production of Vinyon in the United States in 1939. One of the many uses included nonwoven fabric. Vinyon was the bonding agent for the nonwoven fabrics.

1941
SARAN - The Firestone Plastics Company first produced Saran commercially in 1941. The sunlight resistant property made this a very attractive fiber for seat coverings, doll's hair, and outdoor furniture.

Dramatic growth in the use of man-made fibers took place after World War II when new technologies served the needs and lifestyles of the postwar generation.

1946
METALLIC - Three new generic fibers -- metallic, modacrylic and olefin -- were in production by the end of the 1940s. Dow Chemical Co. introduced metallic fibers in 1946. These created luxurious effects for apparel ranging from Paris couture to the American fashion mass market. They have affected the home furnishings textiles as well. Metallic fibers could be plastic-coated metal, metal coated plastic, or a core completely covered by metal. Metallic yarns can be woven, knitted, or applied to fabrics — from the sheer chiffon to tough upholstery.

1949
MODACRYLIC - Introduced by Union Carbide Corporation, the fiber is made from copolymer resins – (combinations) of chemicals derived from natural gas, coal, air, salt, and water. Application of heat produces fibers of different lengths, creating a surface resembling natural fur. Fun furs have been used for items such as apparel, carpets, and blankets.

1949

OLEFIN - Hercules, Inc. developed olefin in 1949. The fiber was structured as a multi-filament for general textile applications by 1961. Resistant to mildew, easier to dye goods, good bulk cover, it found a larger market in home furnishings, especially for indoor and outdoor carpets.

1950

ACRYLIC - DuPont introduced acrylic as a man-made alternative to wool. Its characteristics include softness, warmth, light weight, resiliency, and dye acceptance — enabling it to be used in sweaters and knitwear.

1953

POLYESTER - Although it was first developed by British scientists at the Calico Printers Association in England, DuPont began its production in 1951. The term "wash and wear" was coined in the following year. It was the beginning of an era of easy care apparel — machine washable, easy to dry, crease resistant, retained its shape, and needs no ironing.

1954

TRIACETATE - The Celanese Corporation introduced the triacetate fiber in 1954 in the United States. Triacetate has an excellent pleat and crease ability when set with heat, and it is used in robes, skirts, dresses, and knitted sleepwear.

1960

SPANDEX - Spandex is lightweight, stronger, and more durable than rubber, and it was first produced by DuPont in 1960. It can stretched repeatedly and still recover to its original length. Spandex, therefore, can be found in fashion wherever stretch and freedom of movement is needed, especially from swim wear, dance and active wear, hosiery, and intimate apparel.

1961

ARAMID - DuPont also introduced the Aramid fiber, which is light and extremely strong. It does not melt and is extremely tenacious. Its ability to withstand heat and unusual strength makes it suitable for protective clothing, sporting goods, and bulletproof vests.

1970

ANIDEX - Production of Anidex fiber in the United States began commercially in 1970 by the Rohm and Haas Company. It is used in swim wear and ski wear where stretch is a desirable property.

1974

NOVOLOID - Production of Novoloid fiber began in 1974. Although not produced by the United States, American Kynol, Inc. is the distributor. This fiber is used in protective clothing because of its excellent flame resistance.

Introduction

Fabrics of the 1970s were as distinct as the clothing into which it was made. Bold colors and patterns that dominated fabric design ranged from flamboyant to garish. Dizzying complementary color and pattern combinations, called psychedelic, were adapted from sources as diverse as non-Western folk art and Western technology. Some were inspired designs; some were arbitrary and thoughtless copies of inspired designs. Both represented the decade.

There were also two different categories of fibers used in the 1970s – the naturals and the synthetics. There were wonderful silks, cottons, rayons, and wools, but it was the synthetics – all of those amazing stretch knits of nylon and polyester – that represent the decade. The irony was that the '60s counterculture advocated things like "earth shoes," native folk art, relaxed and/or rustic life style, natural foods, and natural fibers. When their radical clothing styles and colors became more familiar and ultimately more fashionable with the general population, synthetic stretch knits became the rage. Hippies never wore polyester bell-bottomed pants suits, but the rest of the population turned them into a '70s icon. Great and not-so-great textures, metallics, plaids, and prints were all made of synthetic fibers that ranged from the look and feel of silk to the feel and smell of industrial carpet.

The following photographs (mostly close-ups) of fabric swatches cannot be felt. But they certainly can provide a sense of the visual delight (or dismay) that the actual fabrics convey. Each is a miniature historic document that both establishes its place in time and serves as a reference for future designs. The chronology and glossary focus on the man-made, or synthetic, fibers — terms and dates that will aid one in understanding some of the materials, processes, and distinctive qualities that these fabrics possess. The bibliography will lead the reader to more detailed technical, historic, and other information. Our main intent is simply to present a visual sample of these synthetic '70s morsels for the shear pleasure of seeing them.

Solids & Textures

Frost faded leather look of polyurethane for outerwear.
Wamsutta
1974

100% acetate faille taffeta moiré, great for evening wear.
Earl-Glo
1973

The quilted look in pink/white/lavender double knit.
Downen
1970

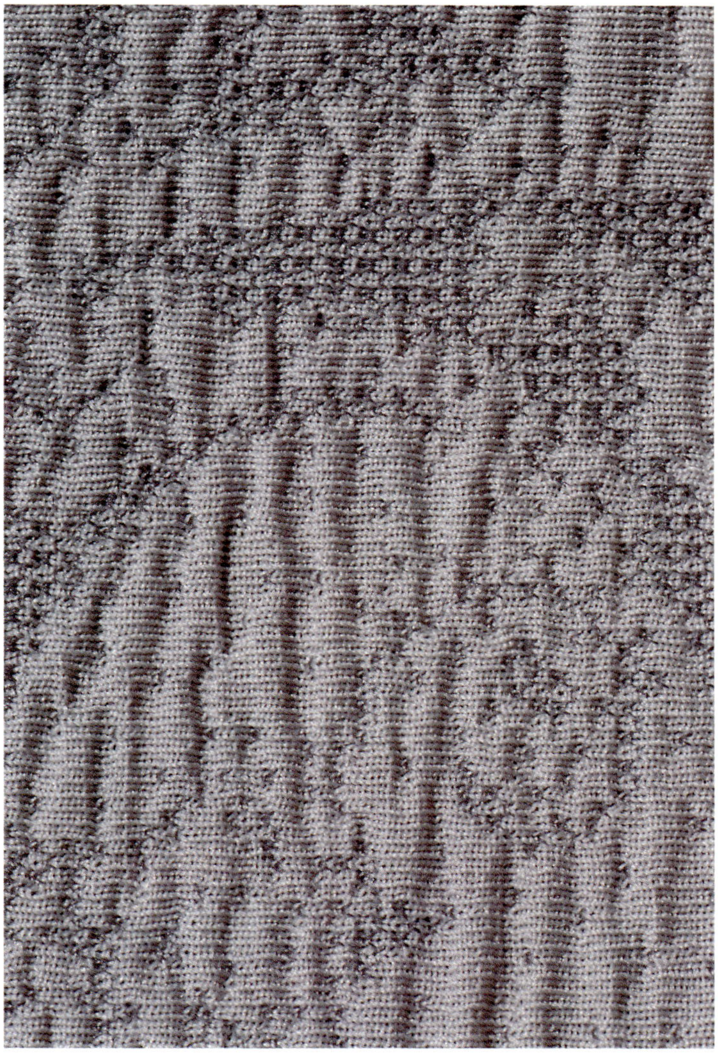

Grey double knit pebble lace look.
Deering Milliken
1970

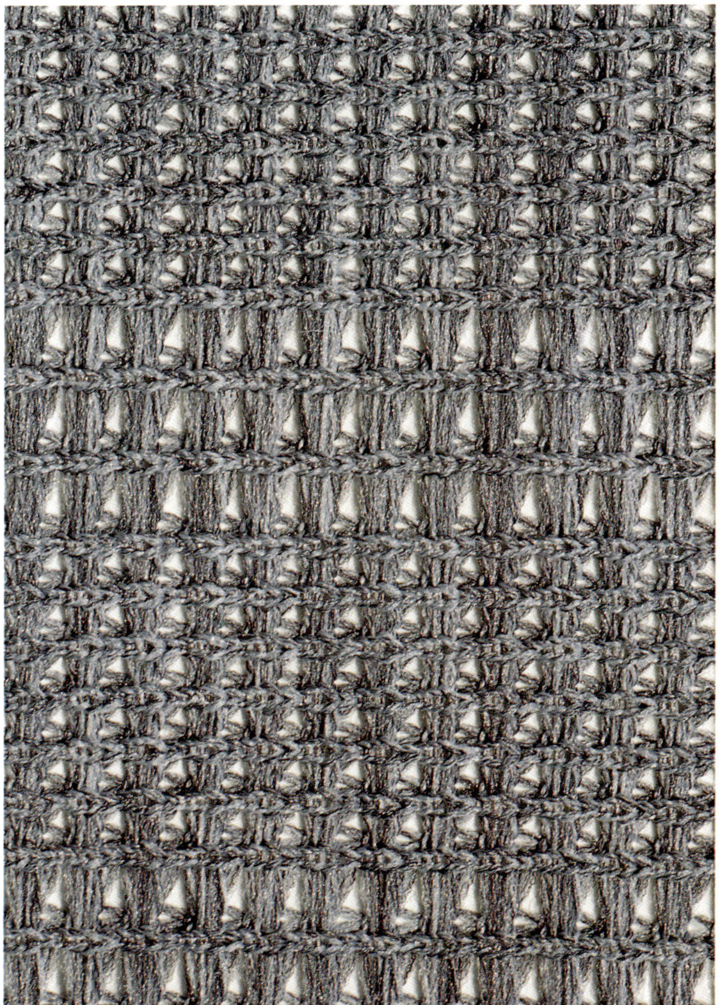

Crochet look lace knit, 100% Fortrel polyester double knit for sweaters and vests.
Winston Mills
1975

Bias-stripe effect double knit blend of Fortrel polyester and wool for men's suits.
Barbet
1971

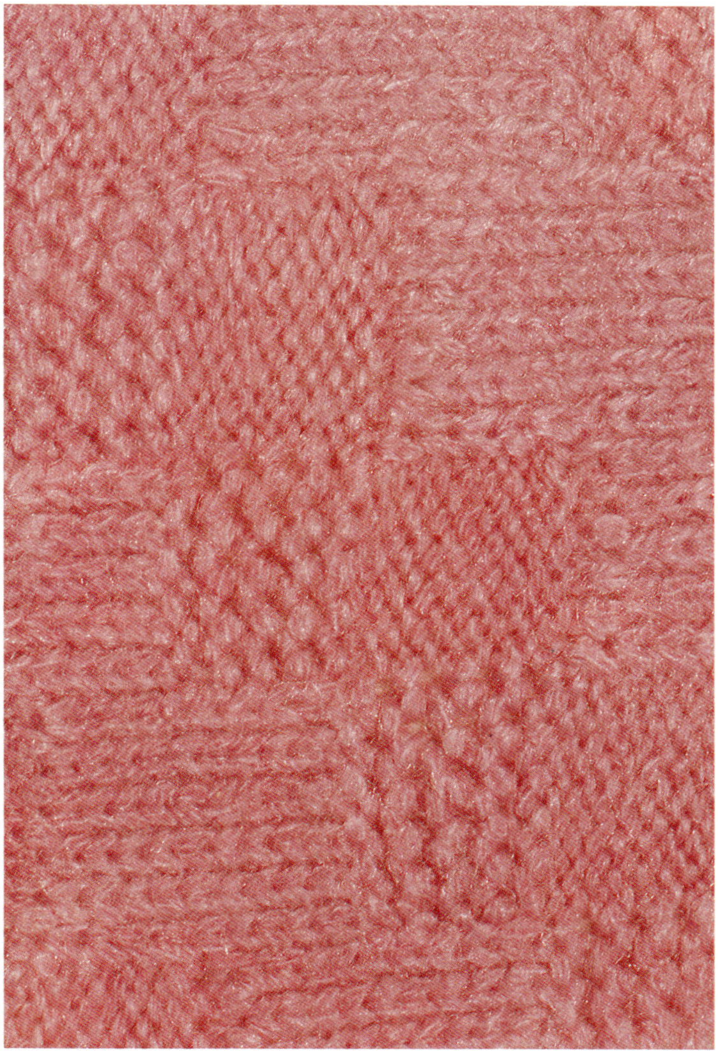

The look and feel of cashmere knit,
warm supple softness, 100% Norvelle
yarns of Orlon acrylic.
DuPont
1978

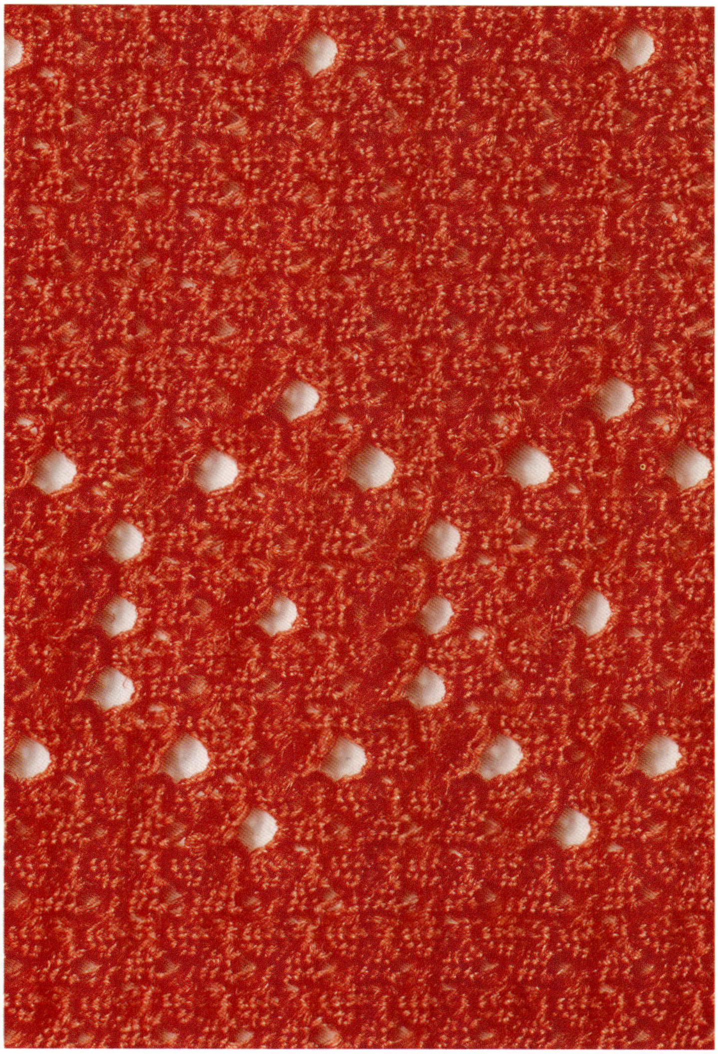

Peekaboo knit 100% Antron nylon,
Persian inspired.
Suemel
1970

Top left: Lacy single knit used for blouses and shirts. West Point Pepperell 1972

Top right: Green striped bouclé rough texture. Galey & Lord 1972

Bottom left: Taupe Silk. 1970

Bottom right: 50% acrylic and 50% rayon, Mexican handicraft look used in men's pants and women's sportswear. Earl-Glo 1971

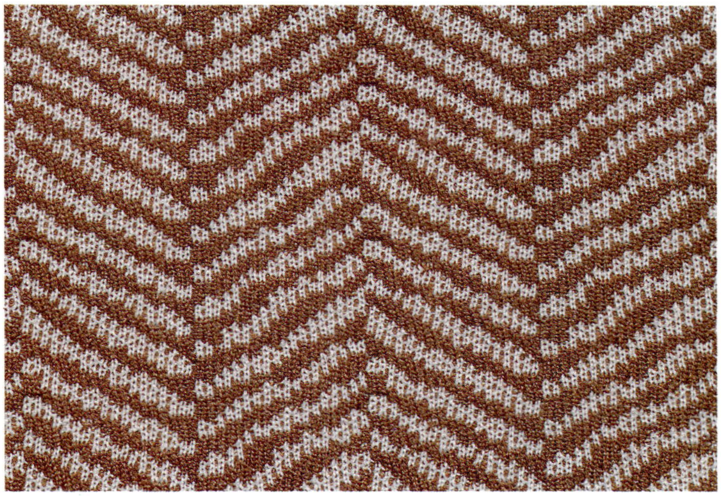

Top: Raised embossed laminated nylon.
Guilford
1972

Bottom: Brown and cream herring-bone men's wear double knit, wrinkle resistant.
Milliken
1971

Top: Uneven texture simulates shantung for separates, slacks, and jackets.
Earl Loom
1971

Bottom: Purple 100% polyester herringbone knit.
Downen
1970

Top: Piece-dyed jacquard double knit from Paris.
Cocitex
1971

Bottom: Single knit Acrilan/linen fabric for sweaters classic white and black.
Scottex Corporation
1973

Top: Textured window pane, fire retardant, clean classic look, 100% polyester.
Kaufmann
1977

Bottom: Yarn-dyed single knit Pre-Columbian inspired 100% Acrilan.
Milliken
1971

Top left: Soft texture basket weave with ribbon embroidery effect in a ribbon three dimensional knit.
Alamac
1977

Bottom left: Integrated print and fabric for skirts, pants, and jackets.
Andrex Division
Duplan
1974

Top right: Earth tones background, raised novelty knit for sportswear.
Shirley Fabrics
1975

Bottom right: Printed chinchilla knit 100% Orlon (face) and 100% cotton (back).
Malden Mills
1972

Brown woven, resists pilling and snagging, used for sportswear.
Swift
1971

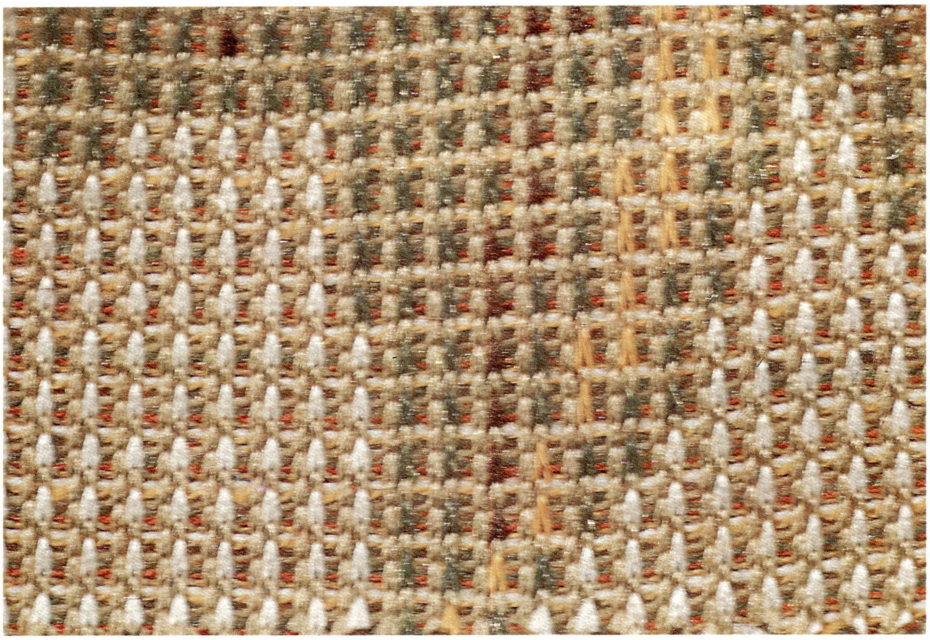

Tapestry knit for upholstery.
Sulzer Bros.
1974

Lush curving large scale floral shapes used in upholstery.
Chromatex
1978

Double knit 100% textured Avlin.
Winner Textiles
1972

Warp knit casement fabric with texture.
Celanese
1974

Stretch fabric with flexibility used for garments for both men and boys.
Swift
1971

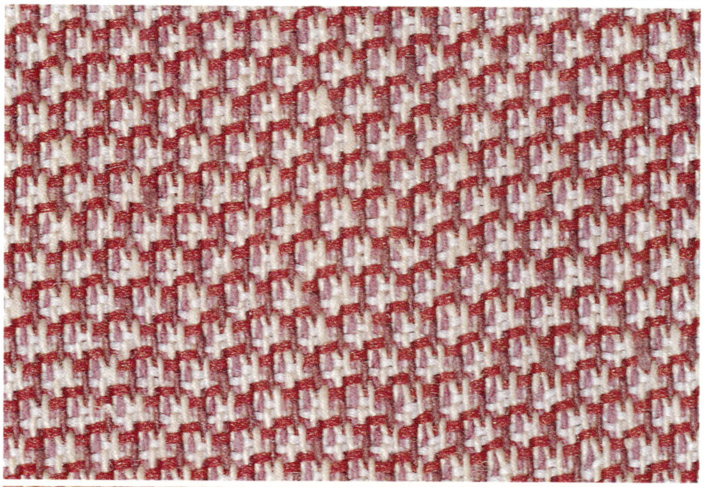

Two-color loose weave blend and machine washable.
Chatham
1971

Top left: Double knit blend of Trevira and silk. ESEMBC Knits 1972

Top right: Brown subtle Raschel fabric made with a rack-type stitch with slubs. Arthur Feinberg Novelty Textiles 1970

Bottom left: Skinny rib hand-made look, 100% textured polyester for sweaters. Cohama 1972

Bottom right: Trevira polyester, lilac and ocher, double knit raised surface brocade look. Downen Knitwear Ltd. 1970

Two tone orange, double knit of whirling twirls.
Deering Milliken
1970

Waffle piqué fabric, 100% textured Dacron yarn.
Cohama
1972

Cotton polyester "little squares" print.
Early 1970s

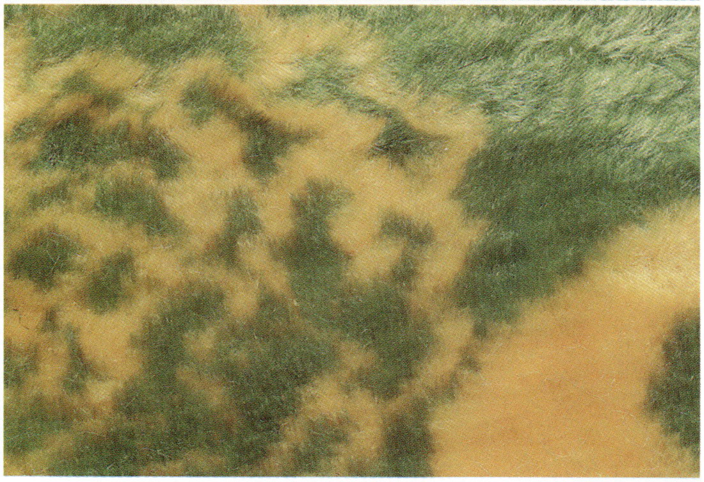

100% acrylic upholstery pile fabric.
Borg
1974

Metallic & Novelty

Knitted wool and metallic glitter trend.
Lebanon Mills
1970

Knitted metallic yarn.
Silan
1971

Silver lining drapery.
Jack Lenor Larsen
1978

Knitted silver and polyester, made on a Raschel machine, designed for jet plane interiors.
Jack Lenor Larsen
1971

Durene, polyester and silk slub sprinkled with gold metallic, a ribbed knit.
Knits, Inc.
1973

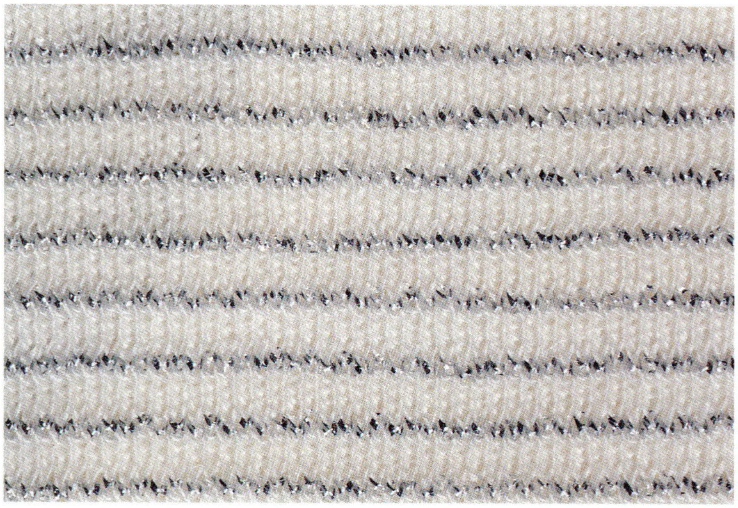

Top: Silver metallic thread on white knit.
1970

Bottom: Acrylic glitter knit with metallic threads woven into the stripes.
Cyanamid
1974

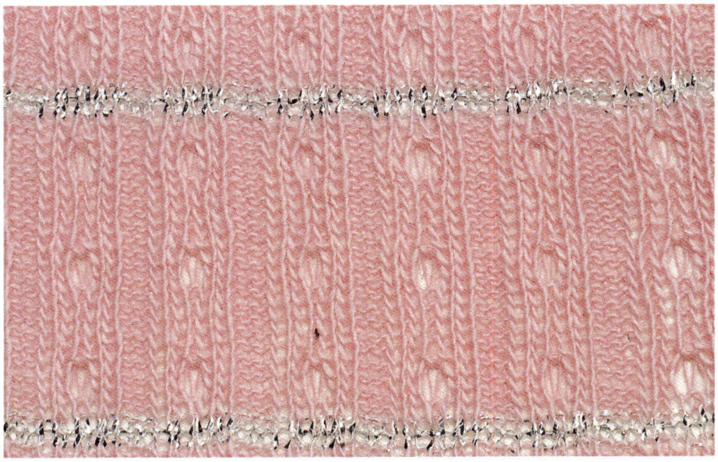

Top: Cotton knit with metallic Lurex stripes of silver on pink.
Marcrest Fabrics
1972

Bottom: Single knit jersey silky soft metallic, great for shimmering nightwear.
Sildorex
1970

Top: Green and gold Acrilan and bright trilobal nylon.
Starward Fabrics Ltd.
1970

Bottom: Knitted metallic yarn.
Silan
1971

Above: Green knit French metallic yarn.
Le Jersey Viennois S.A.
1971

Top right: Aqua knitted metallic yarn.
Le Jersey Viennois S.A.
1971

Right: Metallic yarn does not tarnish, glitters softly.
Sildorex
1971

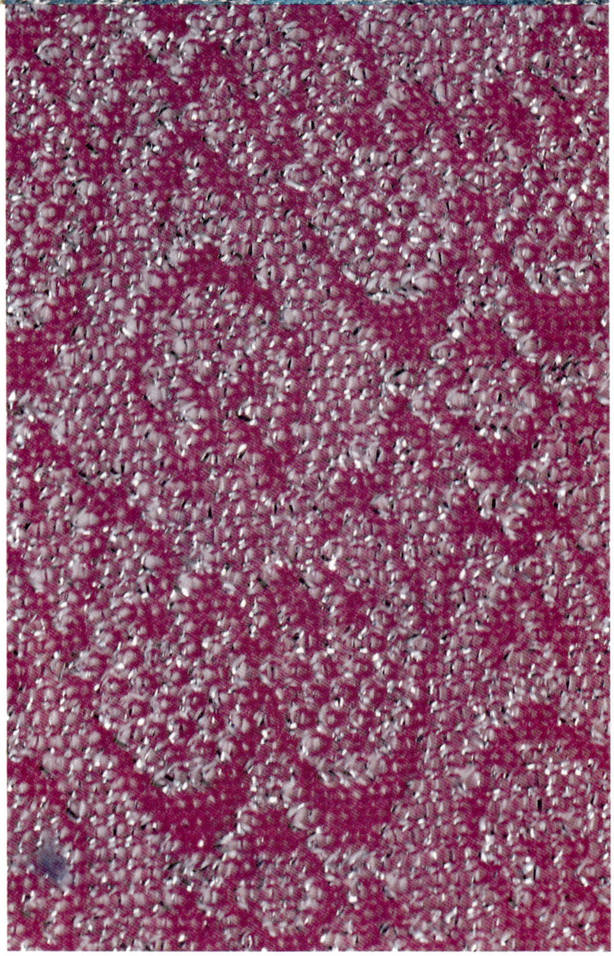

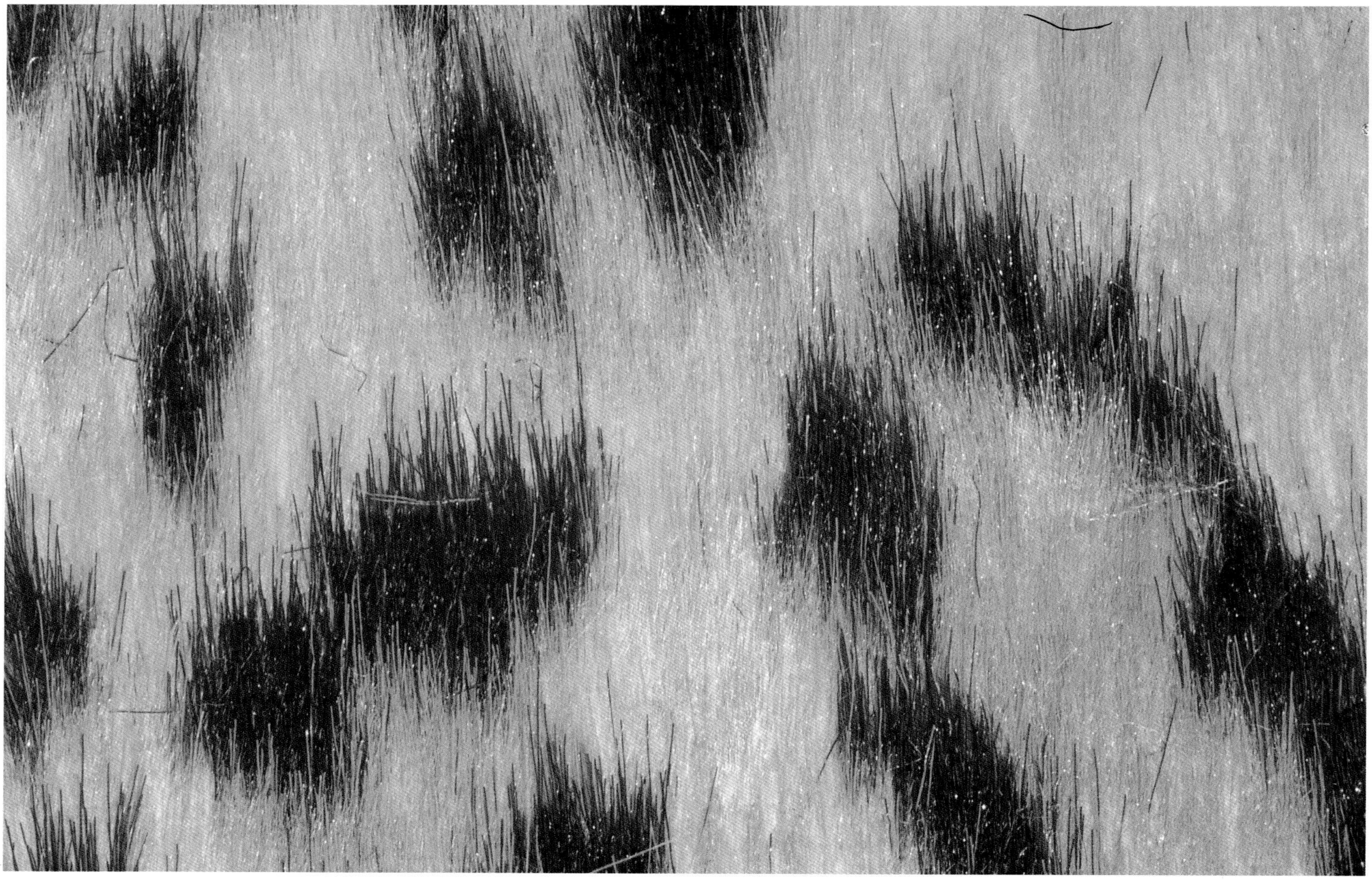

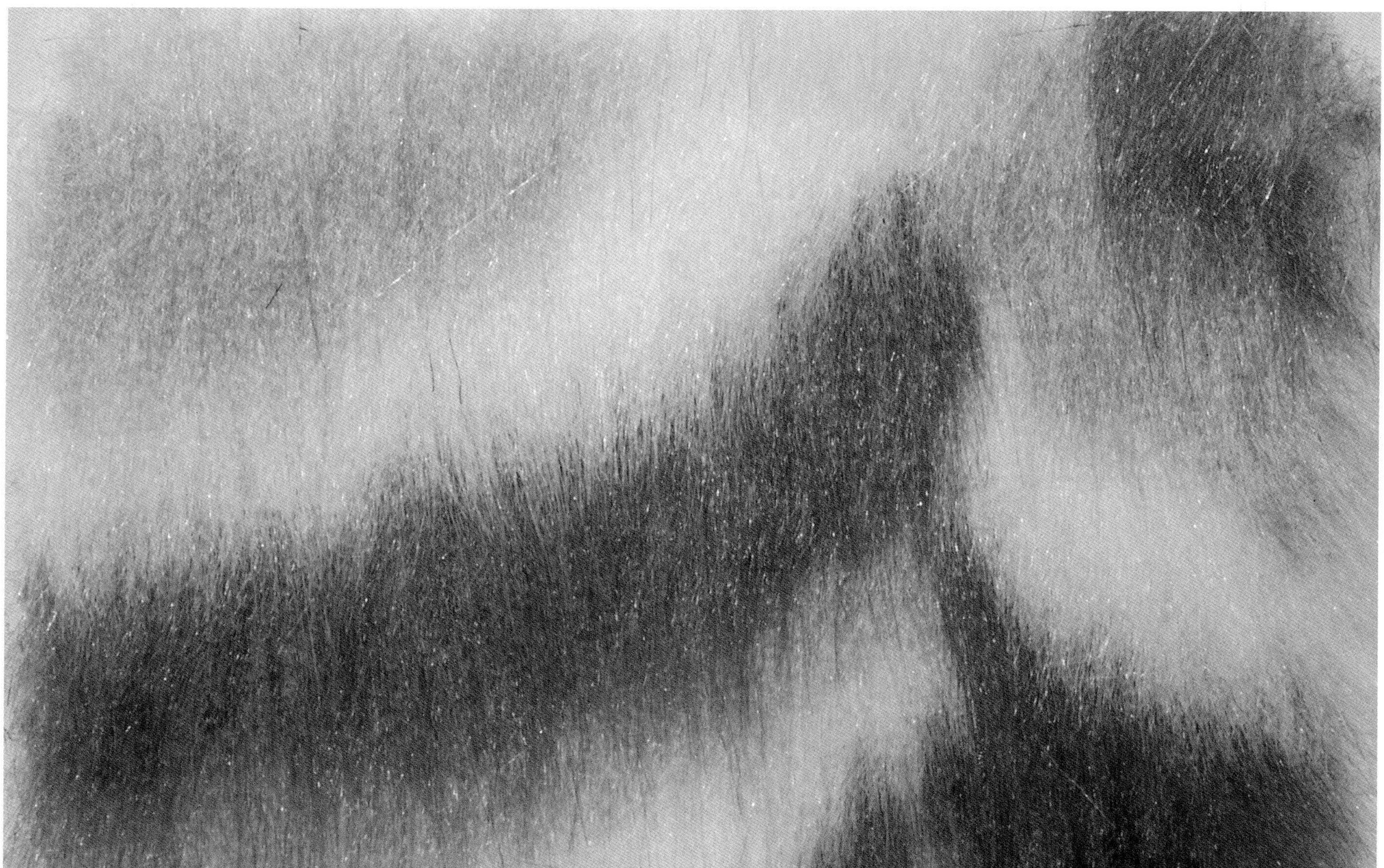

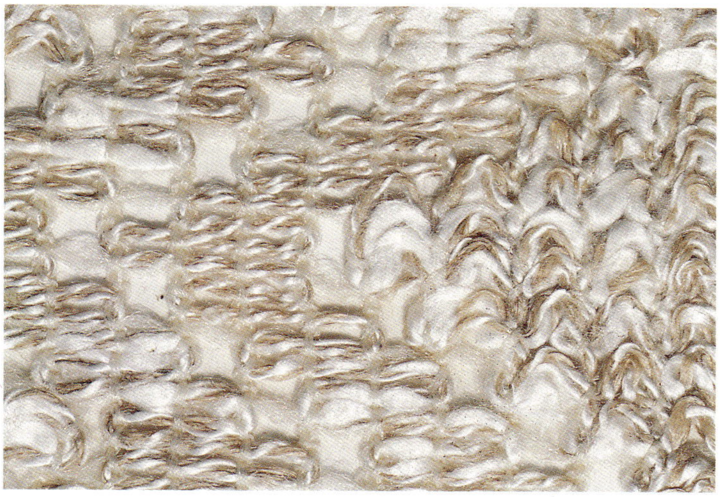

Top: A ribbon weave gives apparel soft hand and surface interest.
Auburn
1971

Bottom: White and beige polyester sweater knit.
Lawson Fabrics
1970

Top: Raschel crocheted look for sweater knits.
Novelty Textiles
1973

Bottom: Dacron, earth tones, stock-dyed, natural look for upholstery.
Guilford
1974

Opposite page:
Top: Exotic blended ocelot print.
Borg
1974

Bottom: Synthetic pieced mink, luxuriously rich fur-pile animal print.
Borg
1974

Above: Stripes, eyelash polyester blend for women's shirts.
Dan River
1972

Top right: Ribbon knit novelty fabric acetate ribbons are knitted together with acrylic yarn.
T. Romer Fabrics
1972

Right: Yarn dyed, casement fabric, machine washable, highly textured, used in home furnishings.
Waumbec
1974

Opposite page:
Top: Dacron, nubby texture weave, natural fabric for upholstery.
DuPont
1974

Bottom: Soft muted color, with open airy weave.
Jack Lenor Larsen
Mid 1970s

Jacquard pattern has the look of hand embroidered needlepoint work.
Milliken
1971

Sweater glitter knit polyester and nylon metallic fabric.
Fabric One Ltd.
1973

100% polyester permanent press with the look of hand crochet work.
Cone Mills
1970

Opposite page:
Top: Casement curtain made on a Raschel machine with an embroidery look to the fabric.
Arlen Fabric
1972

Bottom: Raschel peasant embroidery.
Native Lace
1972

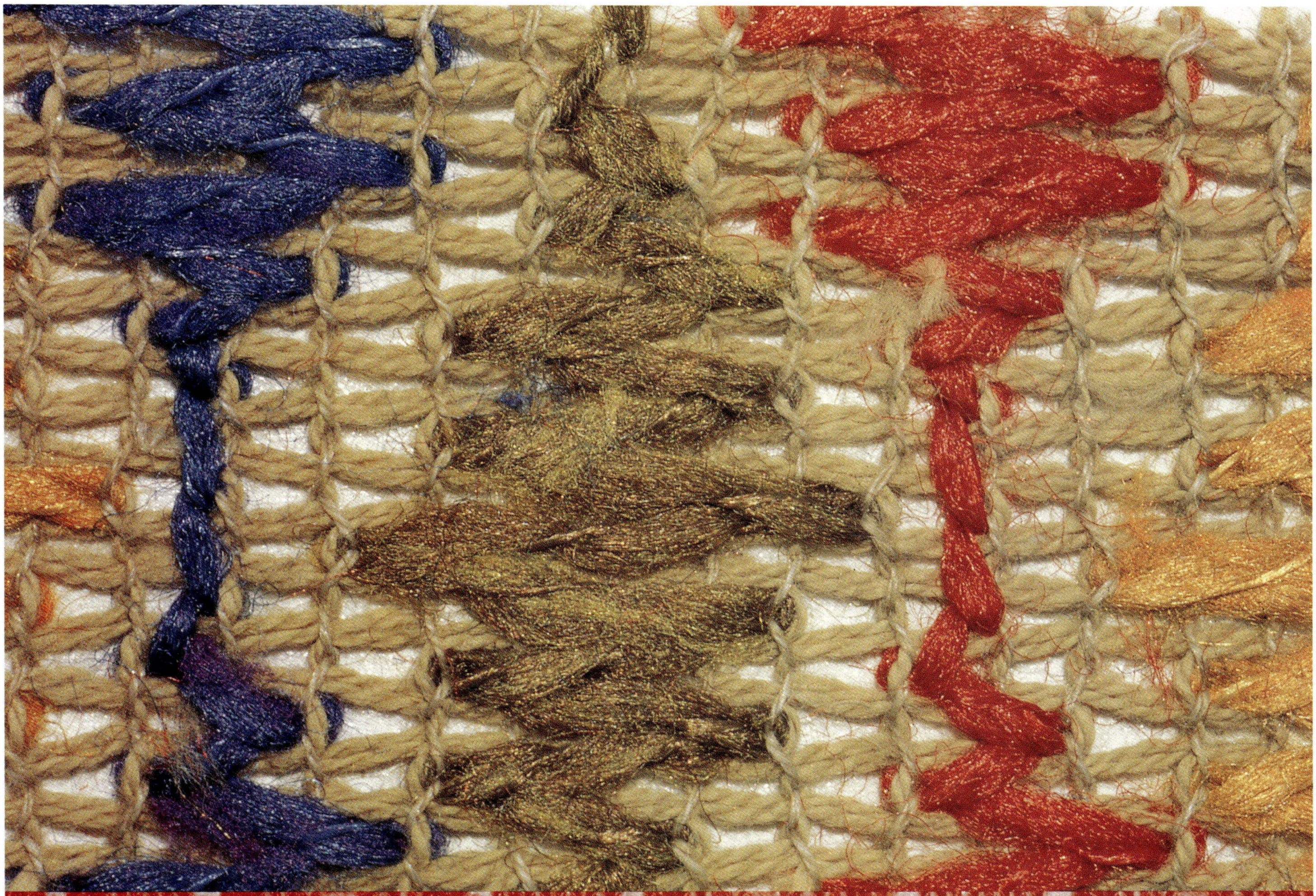

Opposite page: Multi-color flame, open weave knit.
1970

Detail.

Textured open, airy look warp knit drapery fabric.
Guilford
1974

Stripes & Plaids

Colorful slick polyurethane striped surface, weatherproof, light weight, ideal for travel.
Brunswick
1978

Blended, printed, soft gold with a stripe, creates a shiny satin look.
Stanley Looms
1977

Textured illusion striped sheer fabric.
Burlington
1970

Variegated stripe in corded fabric used for unisex pants.
Russell Mills
1970

Pale pastel plaid blend.
Early 1970s

Denim knit polyester easy care striped fabric.
Downen
1970

Bright striped knit.
Wedgwood
1970

Bright seersucker 50/50 polyester blend.
Dan River
1971

Opposite page: Cotton/synthetic blend velour, bold red and blue stripe.
Mid 1970s

Top left: Silky two-bar tricot fabric.
Guilford
1971

Top right: Dan River leads the dramatic shirt revolution of the seventies, creating a new look in men's all synthetic shirting.
Dan River
1970

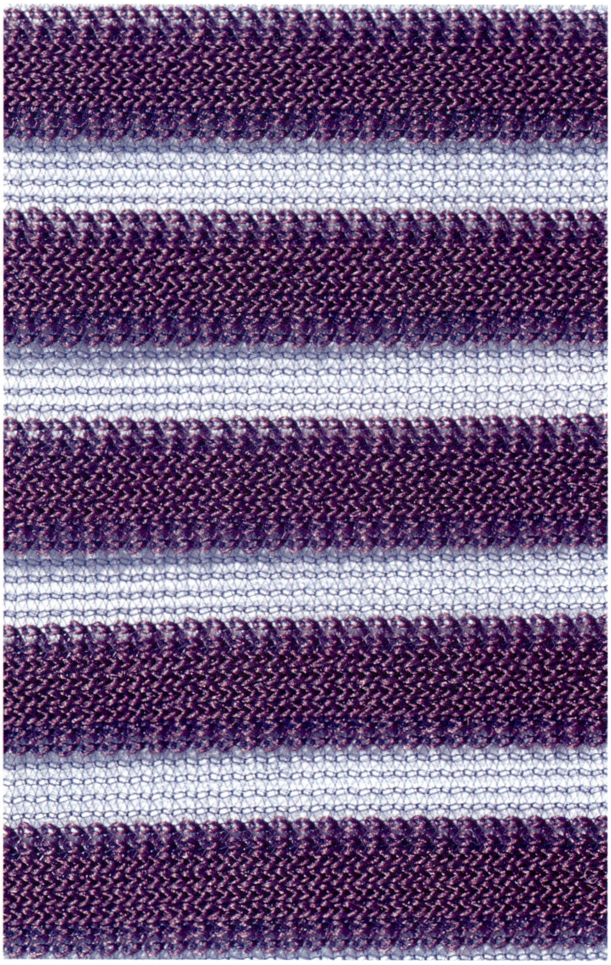

Bottom left: Flash light satin stripe knit, easy care, washable, easy to drape and easy to sew.
Wedgwood
1970

Bottom right: Purple stripe, sheer, knit and unique texture.
Wedgwood
1970

Top left: Cotton knit stretch velour.
Early 1970s

Above: Muted knit, diagonal plaid in wool and Dacron, a Raschel plush fabric.
Knit Couture
1970

Left: Bold stripe, knit and stretches in every direction.
Gilbert & Frank
1972

Opposite page:
Top left: Tartan warp knit 100% textured polyester.
Calina
1972

Top right: Green knit shirting fabric.
Camber
1972

Bottom left: Muted broken plaid textured polyester.
Hoechst and Stehli
1974

Bottom right: Black and red challis hand fashion knit, Arnel blend for blouses and sportswear.
Wullschleger
1974

Opposite page:
Top left: Bulky bold check acrylic bonded fabric.
Herlinger-Bristol
1971

Top right: Plaid wool blend.
J.P. Stevens & Co.
1972

Bottom left: Contrast stitching on denim.
Swift
1972

Bottom right: Permanent press fabric.
Mooresville Mills
1970

Top: Blue knit 100% polyester.
Hoechst
1972

Bottom: Creslan acrylic and polyester double knit used in sportswear market.
J.P. Stevens & Co.
1972

Top: Green plaid knit and 100% polyester.
Hoechst
1972

Bottom: White, brown, black 100% Orlon acrylic bonded to 100% acetate tricot with sculptured chains which cross each other in the middle of the fabric.
Dan River
1970

Top: Black, ivory and beige woven cotton/polyester for women's wear.
Galey & Lord
1973

Bottom: Polyester blend seersucker.
Westwood
1972

Top: Wool and acrylic plaid.
Mid 1970s

Bottom: Permanent press for appearance retention and easy care.
Mooresville Mills
1970

Top: Brown acrylic glen plaid
washable, permanent press.
N. Erlanger Blumgart
1970

Bottom: Reversible checkerboard,
100% Acrilan.
Milliken
1972

Top: Double knit Madras inspired
100% polyester.
Calvert Knitting
1970

Bottom: Permanent press easy-care
plaid.
Mooresville Mills
1970

Madras blended pattern.
Cone Mills
1972

Navy, red, white bold stripes, single knit for men's and women's sportswear.
Caro-Knit
1973

Striped 100% yarn dyed polyester, single knit.
Texfi
1973

Green, navy, and red rayon taffeta.
Mid 1970s

50% Kodel polyester and 50% cotton which has the look of wool in soft colors for fashion in un-denim.
Swift
1970

Great color trompe l'oeil seersucker, crisp, crunchy hand effect of 100% polyester.
Wamsutta
1976

Opposite page: Shagbark cloth of blended polyester.
Galey & Lord
1972

100% blue and gray polyester knit.
Hoechst
1972

Checkmate blend.
Graniteville's Kenville
1972

Brown honeycomb upholstery, deep textured yarn.
Celanese
1974

Polyester and nylon blend.
Graniteville's Kenville
1972

Luxurious hand, locked-in color brilliance, acetate and polyester used in evening wear and ties.
Amer-Mill
1972

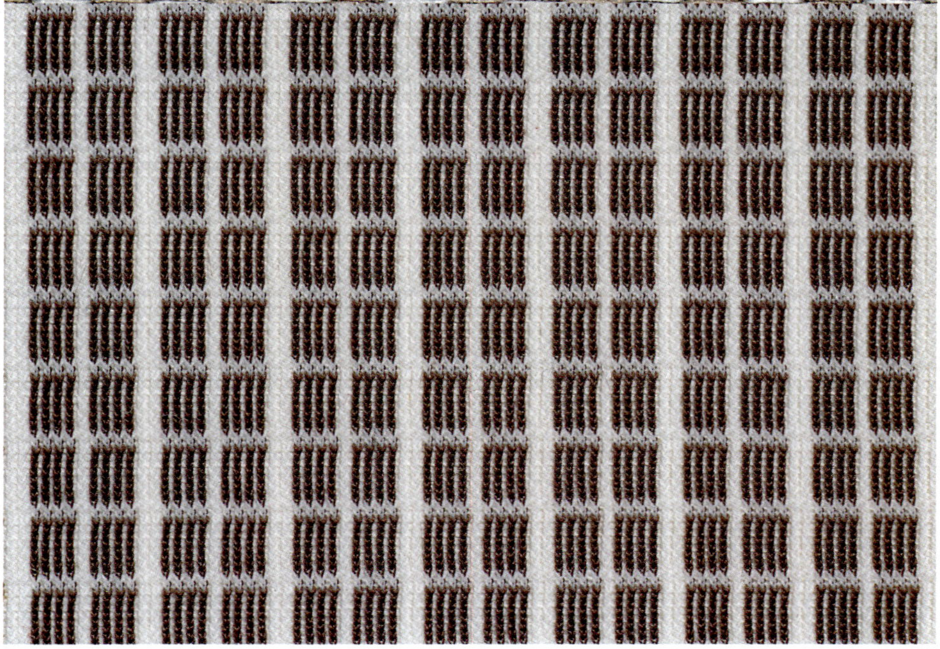

Knit easy-care, washable, drapable, with checks.
Wedgwood
1970

Geometric

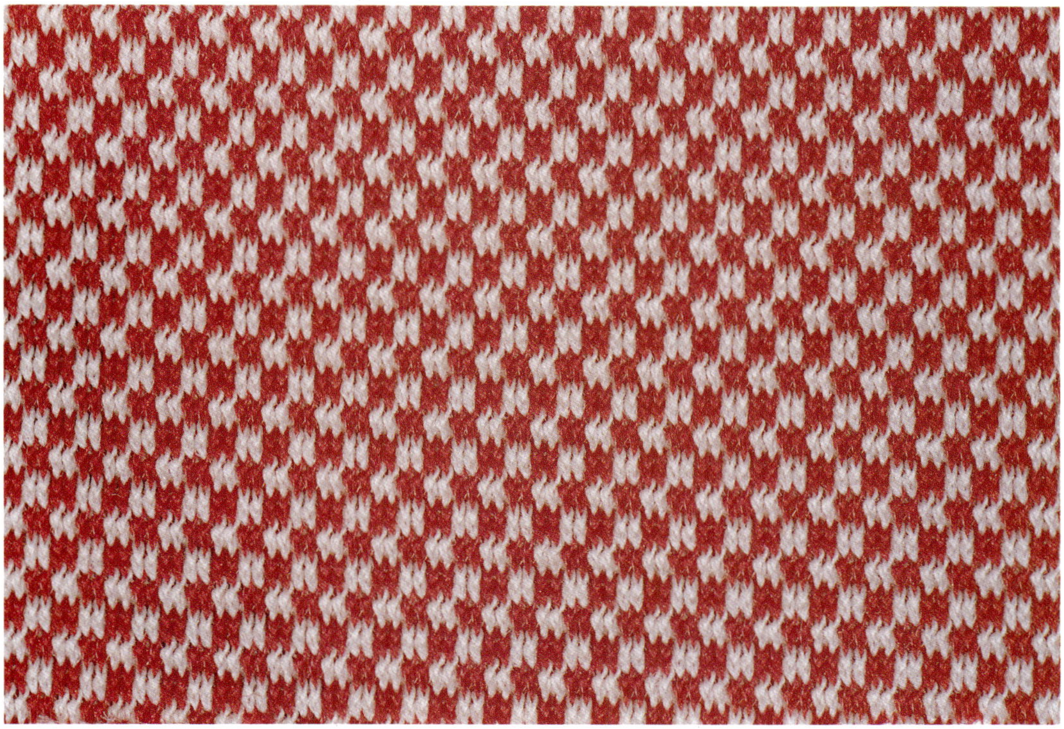

Red and white polyester warp knit.
Guilford 1972

Non-raveling knit, easy care, washable, drape, easy to sew.
Wedgwood 1970

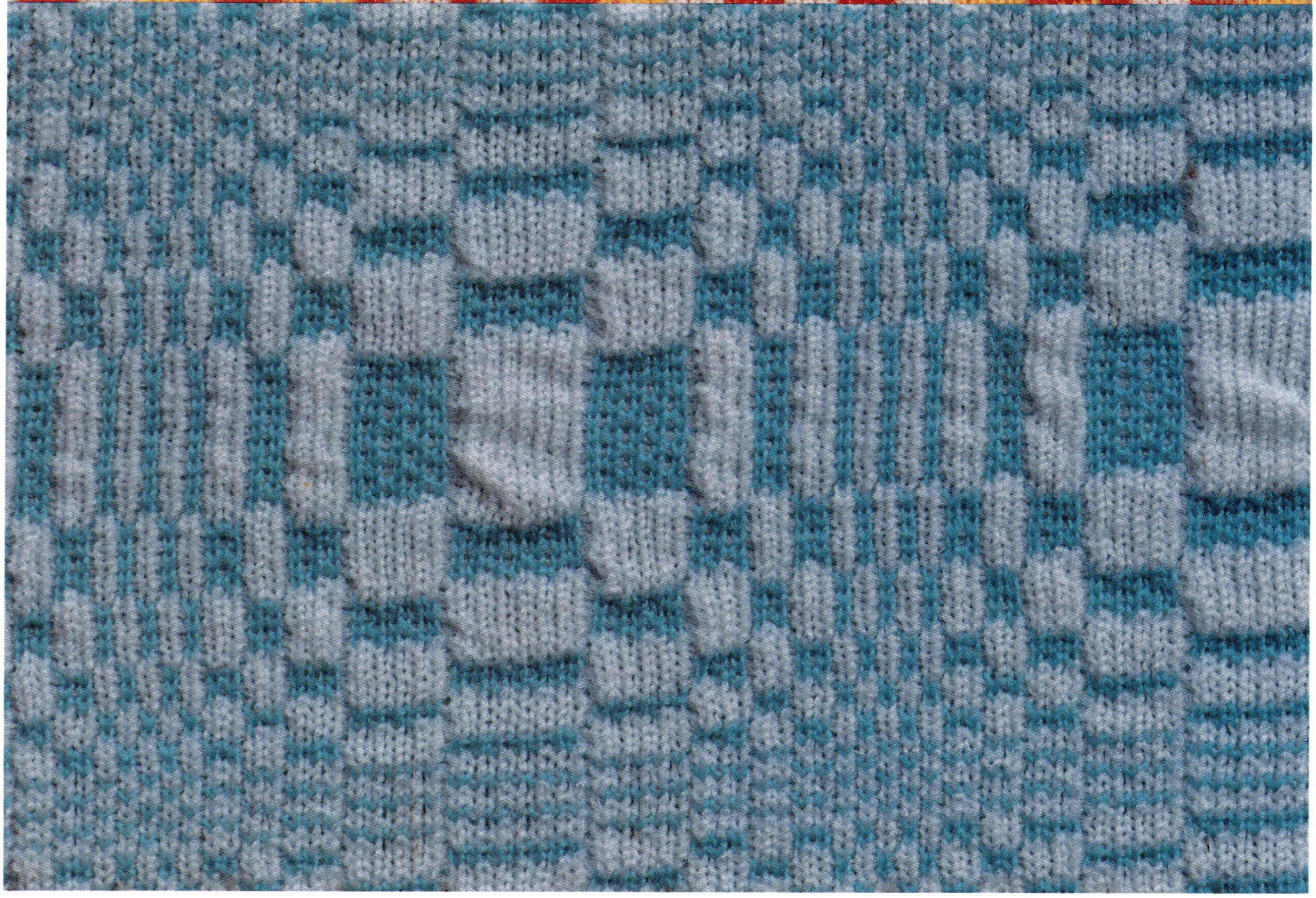

Top: 100% yarn-dyed polyester, jacquard knit, sold in one-stop centers by Singer.
1972

Bottom: Trompe l'oeil 100% Dacron polyester, optical illusion, machine washable, no-iron permanent press.
Deering Milliken
1970

Top: Black and white harlequin pattern.
Late 1970s

Bottom: Trompe l'oeil in 100% Dacron knit, optical illusion of boxes running in three different directions.
Stehli
1970

Top left: Multi-color stretch polyester knit.
Mid 1970s

Top right: Jacquard double knit for women's wear, 100% textured polyester.
Hoechst
1972

Bottom: Red 100% polyester double knit for women's wear.
Texfi
1973

Diamond quilt, double knit blend, geometric, large and bold.
Deering Milliken
1970

Jacquard geometric nylon knit for men's shirts.
Bentley Machinery
1972

Bonded acetate chevron design.
Courtaulds
1972

Men's wear, chevron, double knit blend and wrinkle resistant.
Milliken
1971

Opposite page: Antron nylon print, double knit polyester.
Don-nel Fabrics Inc.
1974

Creslan acrylic and polyester double knit for sportswear and dresses.
J.P. Stevens & Co.
1972

Diamond pattern 100% polyester double knit.
Texfi
1973

Chevron polyester knit.
Guilford
1972

Swimsuit, Lycra with stretch chevron design.
Mid 1970s

Top: Indian texture, polyester/spandex double knit for men's and women's swim wear.
Rohm and Haas
1973

Bottom left: Jacquard upholstery, excellent strength, wrinkle resistant, easy to clean Kodel polyester.
Mastercraft
1977

Bottom center: Inspired from an American Indian blanket motif. Wool and nylon blend, soft, wearable, durable, and strong.
J.P. Stevens & Co.
1970

Bottom right: Knitting with French metallic transparent yarn.
Silan
1971

Opposite page:
Top: Bright colored polyester blend patchwork used for draperies and bedspreads.
A.W. Scheffres
1971
Bottom left: Blend of Fortrel polyester, bold Persian pattern with vertical diamond geometric pattern.
Concord
1970
Bottom right: Geometric polyester knit.
Late 1970s

100% stretch double knit for men and boy's swim wear and women's sportswear.
Rohm and Haas
1973

Variegated geometric stripes and diminishing diamonds, 100% polyester double knit.
J.P. Stevens & Co.
1970

Quilted double knit geometric, 100% Dacron polyester.
Stehli
1971

Jacquard stripe, 100% textured Fortrel polyester for men's slacks.
Barbet
1970

Acrylic jacquard knit, vibrant color in junior sportswear for sweaters.
Marcrest Knitting Mills
1973

Jacquard double knit for women's wear, 100% textured polyester.
Hoechst
1972

Top: Raschel stretch lace, 100% nylon for blouses, dresses, and lounge wear.
Paris Lace
1973

Bottom: Nylon delicate diamond pattern used for blouses and lingerie.
Loom-Tex
1971

Top: Jacquard double knit, embroidery texture, polyester.
Downen
1971

Bottom: Persian inspired double knit textured polyester.
Stretchnit
1970

Top: Single knit, jacquard, blend.
Acadia
1972

Bottom: Jacquard double knit for women's wear, 100% textured polyester.
Hoechst
1972

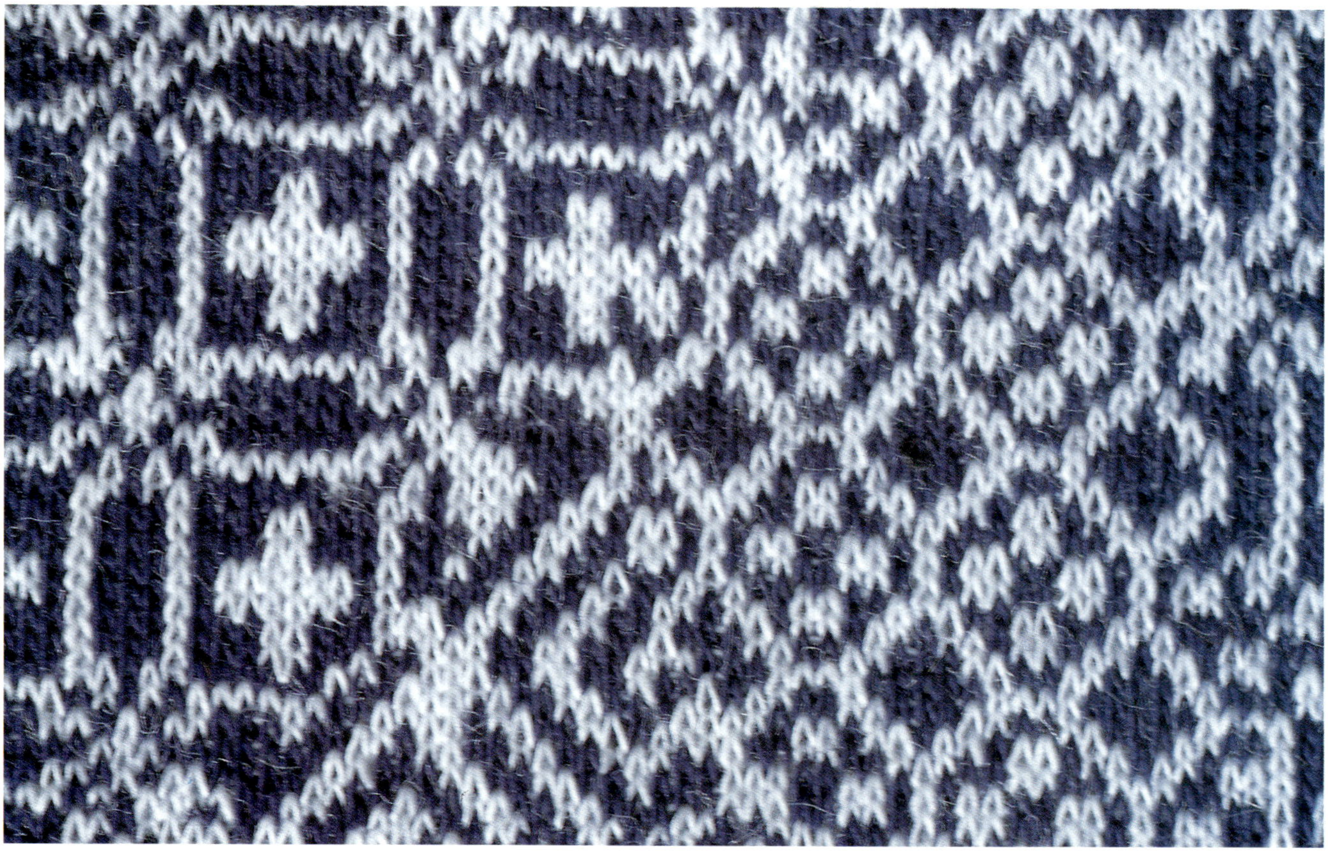

Top: Knit oriental patchwork, double knit of 100% cotton blend in French navy and white color.
Knitbrook Mills
1970

Bottom: The look of tapestry, bright bold geometric squares, inspired by a mosaic plate from Alhambra designed by Robert Reed.
Somerset Mills
1970

Blend, soft subtle geometric design in men's shirting.
Dan River
1970

Two tone orange double knit, geometric inspired.
Deering Milliken
1970

Dacron polyester, two-way woven stretch jacquard.
DuPont
1972

Blend, geometric shapes on a quilt like ground for upholstery.
Collins & Aikman
1978

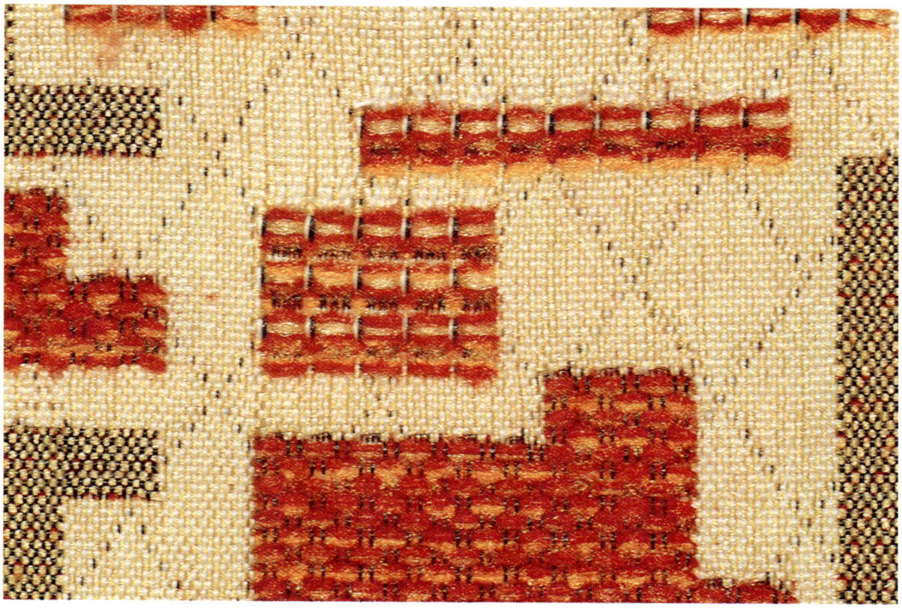

Pastel brights, warp knit blend, print.
Kanebo
1970

Printed warp knit, 100% Enkalure, bright primary colors, geometric, shiny and supple.
Edwards
1973

Top: Printed twill, Qiana for dresses and separates.
Couleur
1974

Bottom: Double knit, graph paper floral.
Chelsea Industries
1972

Top: Interlock fabric, 100% polyester for dresses and sportswear.
Texfi
1973

Bottom: Qiana jersey, closest thing to silk, silky drape for men's shirts.
Couleur
1974

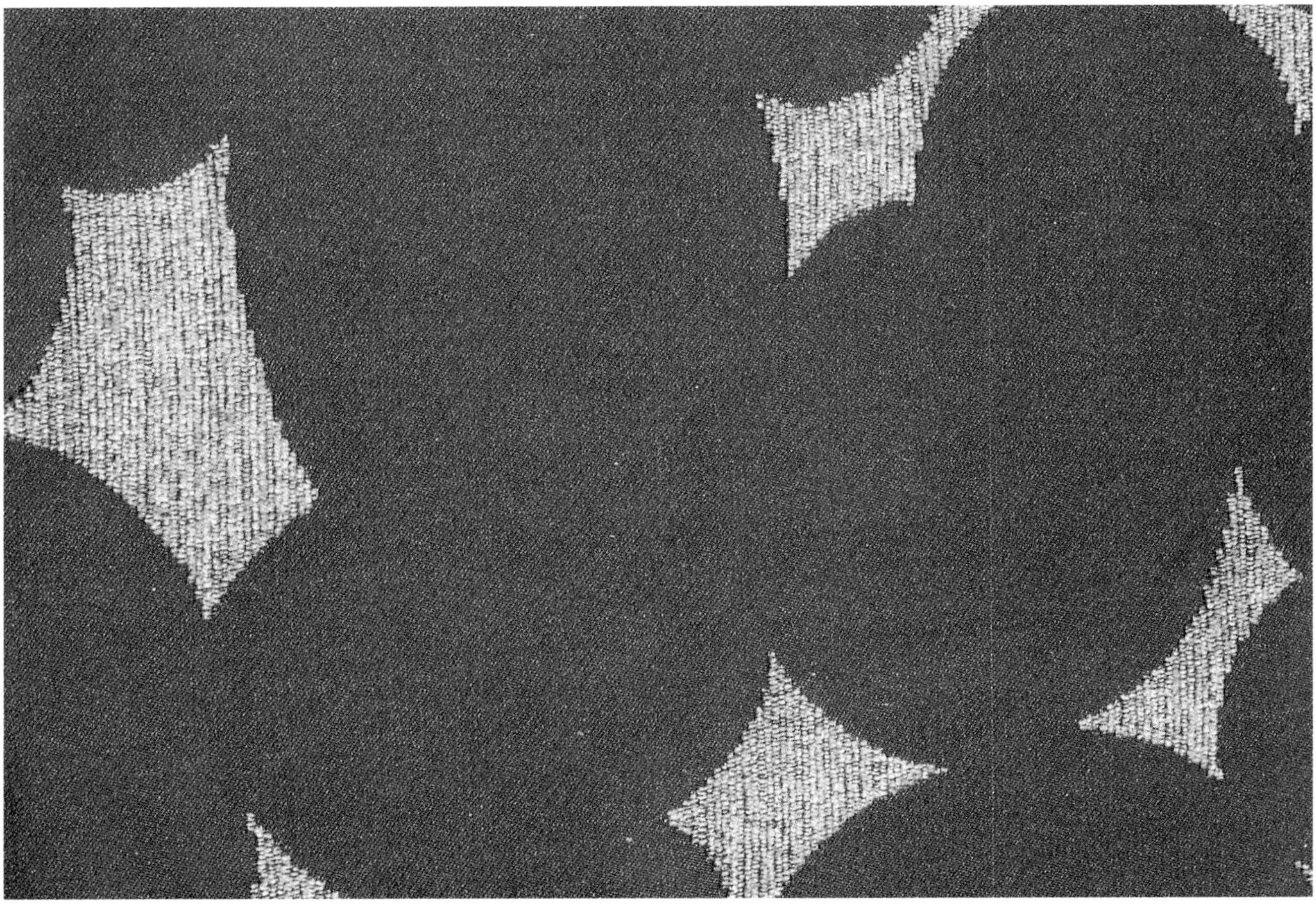

Top: *Stones,* a Wolf Bauer design,
54% cotton, 46% polyester used for
upholstery.
Knoll Textiles
1970

Bottom: Another color variation.
Knoll Textiles
1970

Top: Slinky lightweight Surah blended fabric for women's blouses and dresses.
Renee
1970

Bottom: Tricot polyester, printed green dot knit.
Wedgwood
1975

Bold pattern double knit, 100% textured Fortrel polyester, striking jacquard design.
1971

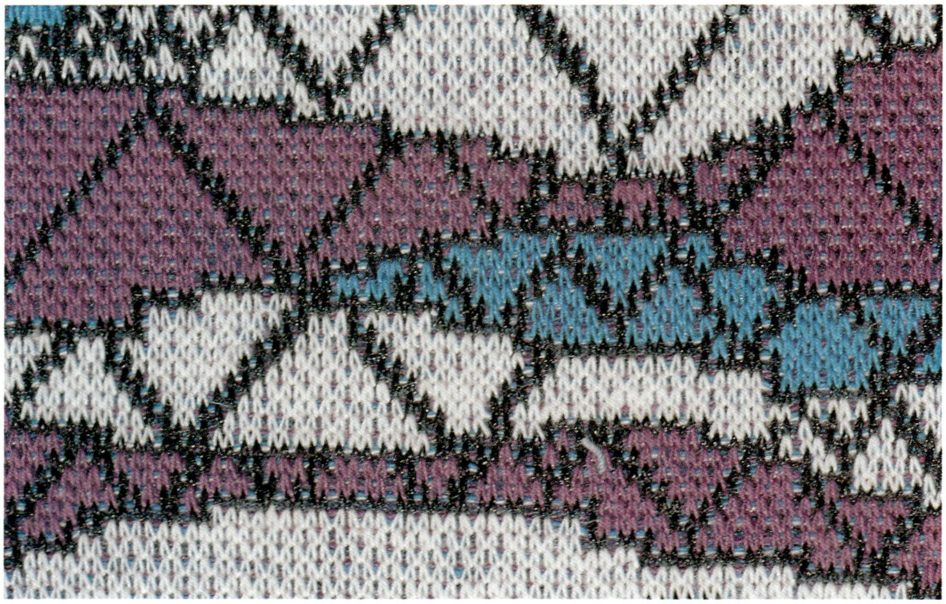

Cotton polyester blended double knit print.
Cotton Incorporated
1973

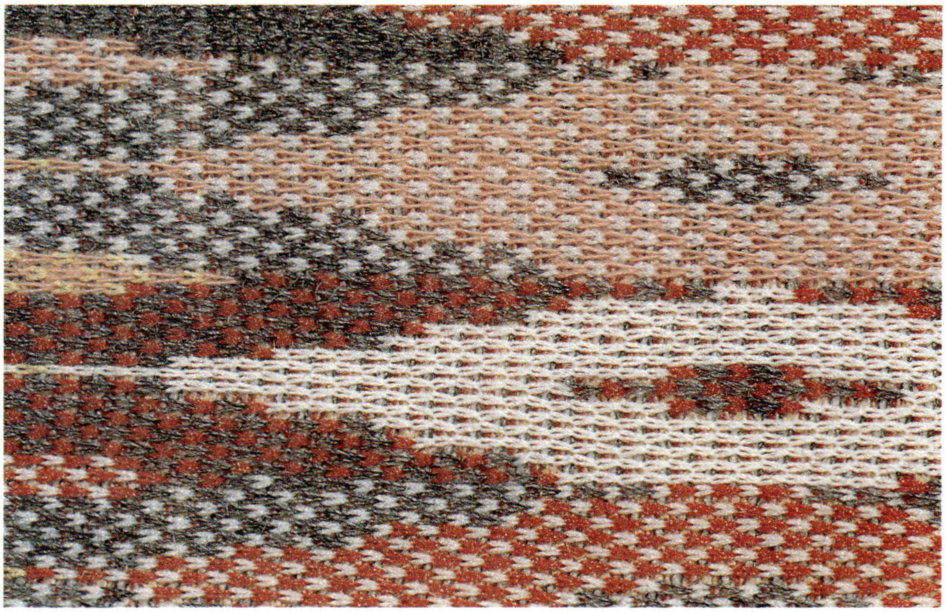

Double knit polyester and acrylic for women's wear.
Texfi
1973

Opposite page: Pucci, knock-off print.
Early 1970s

Abstract

Knit casement, 100% polyester, leafy lace design for a beautiful window treatment, open without transparency.
Belding Hausman
1977

Trevira polyester/ rayon blend print.
Reliable Textiles
1974

Polyester, natural orange, fresh clean floral for draperies and window treatments.
Century Curtain
1977

One color stylized leaf pattern for upholstery.
Cohama
1978

Orange, 100% polyester double knit.
Cohama
1971

Orange, double knit for dresses and sportswear.
Cornell Knits
1973

100% polyester double knit, abstract design.
Soptra
1970

Avril rayon blend, knitted print.
American Viscose
1971

Two-way stretch knit, dress shirt print.
Norwood
1971

Tooled leather look in inlaid vinyl, used in midi vests and gauchos.
Pacific Mills
1970

Opposite page: Red sheer polyester knit.
Mid 1970s

Top: Luxurious Qiana jersey, mixes color with floral for pajamas, dresses, and skirts.
Couleur
1974

Bottom: Nyesta crepe, stretch, slinky, sexy shimmering print.
Don-nel Fabrics Inc.
1973

Opposite page: Flawless print blend, silky hand, premier drape and superior wearable print.
Falcon Mills
1970

Top left: Challis crepe blend, soft stylized floral on a black background.
Cranston
1979

Top right: Silky hand blend, premier drape and excellent wearable print.
Falcon Mills
1970

Bottom left: Arnel jersey knit exotic print, cascade of distorted geometric shapes.
Joyce Fabrics
1970

Bottom right: Nylon silky smooth multi-filament with spandex knit for dresses and blouses.
Charbert
1975

Above: A printed Ban-Lon knit, white, gold, and red design.
Trend Textiles
1970

Left: Green and turquoise double knit, paisley inspired.
Milliken
1970

Top: Geometric polyester single knit, inspired by Persian design of swirl motif.
Cohama
1970

Bottom: Orange and pink double knit, floral print.
Deering Milliken
1970

Top: Jacquard knit in muted colors.
Duplan
1971

Bottom: Irregular geometric, 100% polyester double knit, suitable for sportswear and dresses.
Texfi
1971

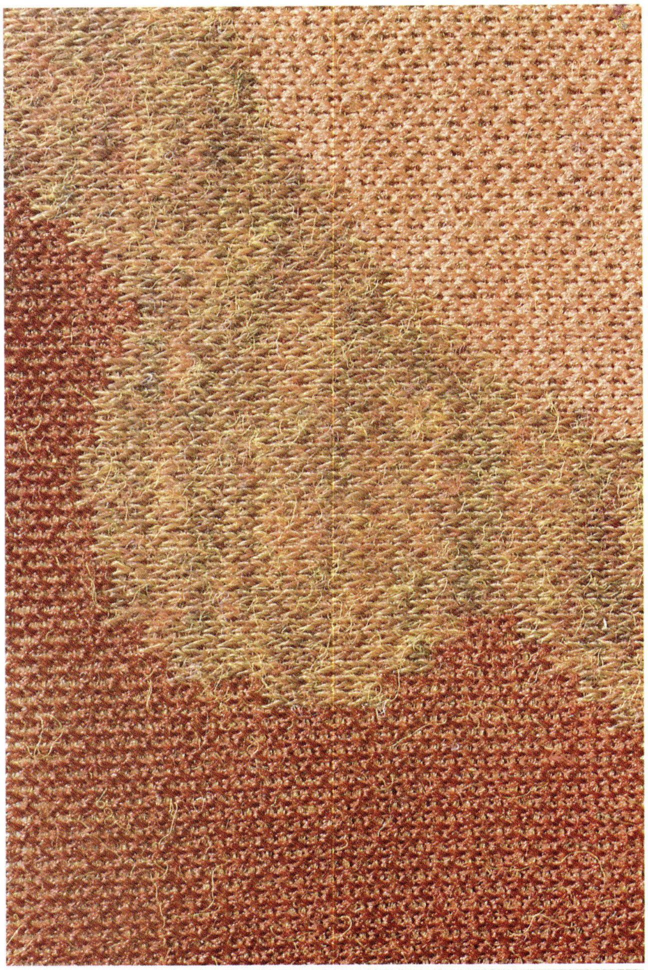
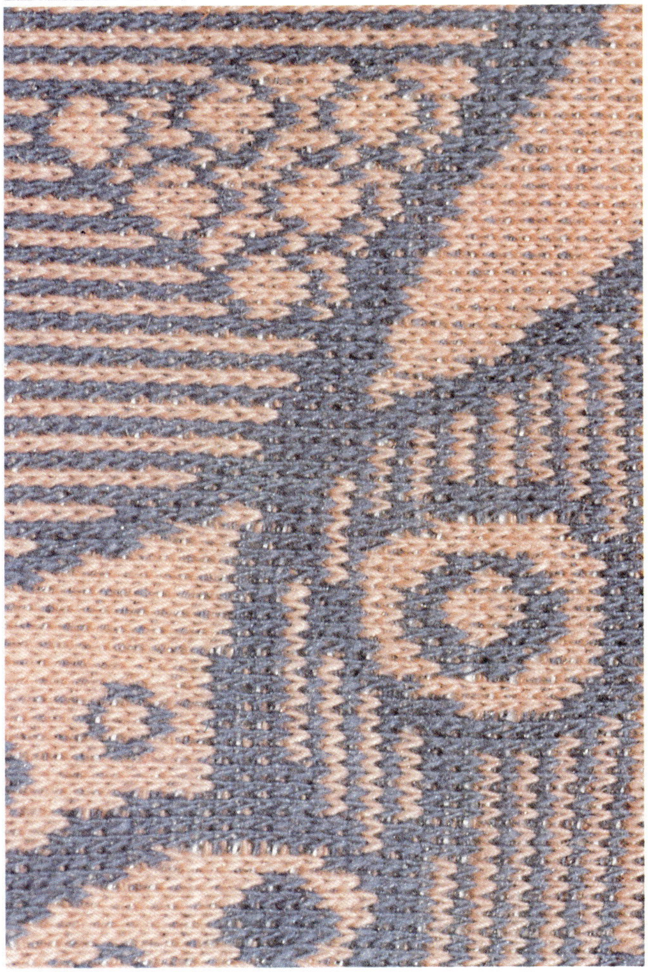
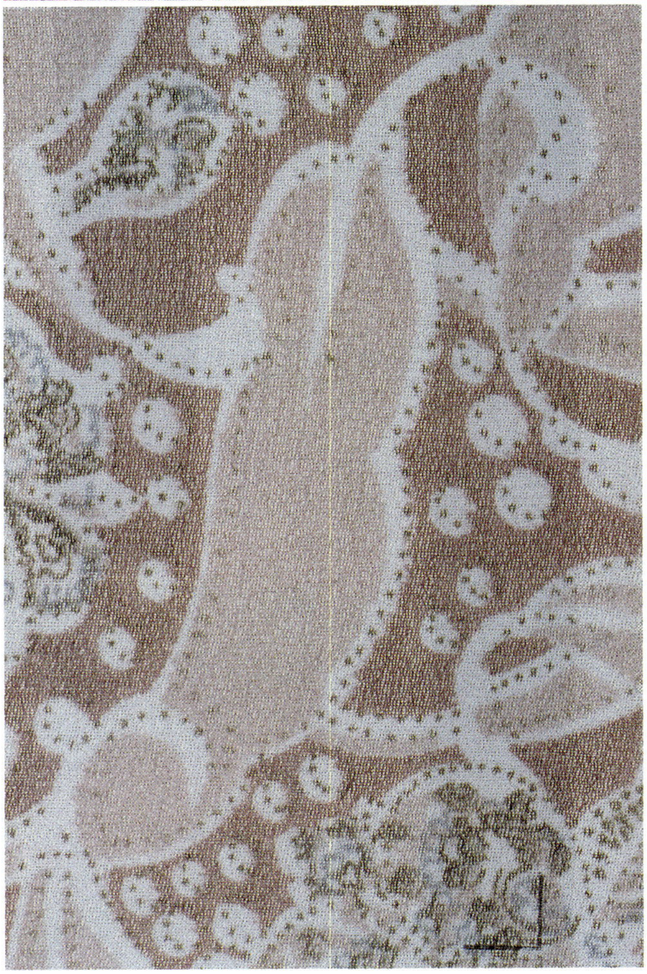

Opposite page:
Top left: Polyester/nylon and metallic double knit for women's evening wear.
Rohm and Haas
1973

Top right: Double knit, piece-dyeing produces two solid colors and a soft field of heather.
American Viscose
1970

Bottom left: Floral double knit cotton and polyester blend.
Cotton Incorporated
1974

Bottom right: Muted tones very sheer nylon stylized flower shapes.
Kenmill
1975

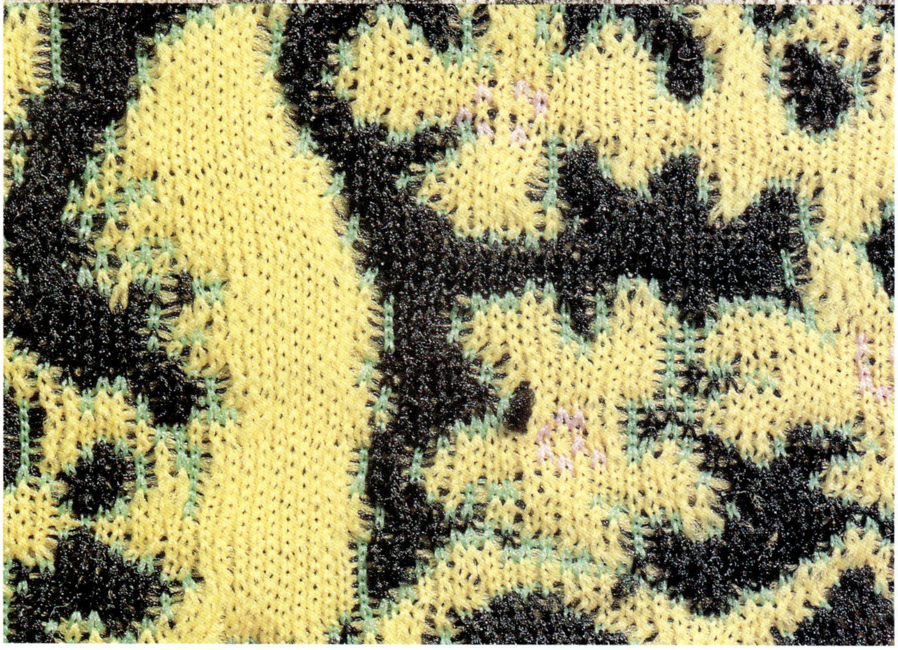

Top: Jacquard double knit for men's wear, 100% textured polyester.
Hoechst
1972

Center: Jacquard double knit for men's wear, 100% textured polyester.
Hoechst
1971

Bottom: Double knit tapestry, Persian inspired, suggests minarets and mosques.
Siltex
1970

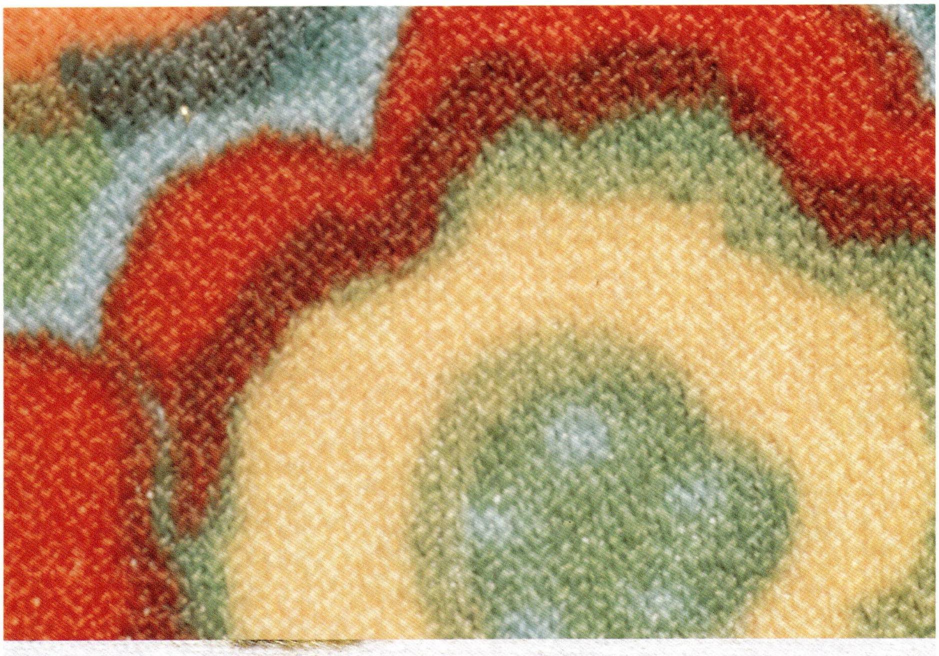

Top: Bold texture polyester and nylon warp knit for women's leisure wear and sportswear.
Rohm and Haas
1973

Center: Laminated velvet nylon print.
Guilford
1972

Bottom: Green/pink/white colorful nylon print.
Courtaulds
1971

Opposite page:
Top left: Cotton and acrylic single knit print.
Cotton Incorporated
1973

Top right: Velvet upholstery in printed Antrol stain-resistant water repellent fabric.
Malden Mills
1974

Bottom left: A kaleidoscope flow of colors, Scandinavian motif in Ombré metallic with a look of psychedelic stained glass.
Kendale Fabrics
1976

Bottom right: Woven polyester and cotton, combines two trends, gauze and ikat.
Ameritex
1975

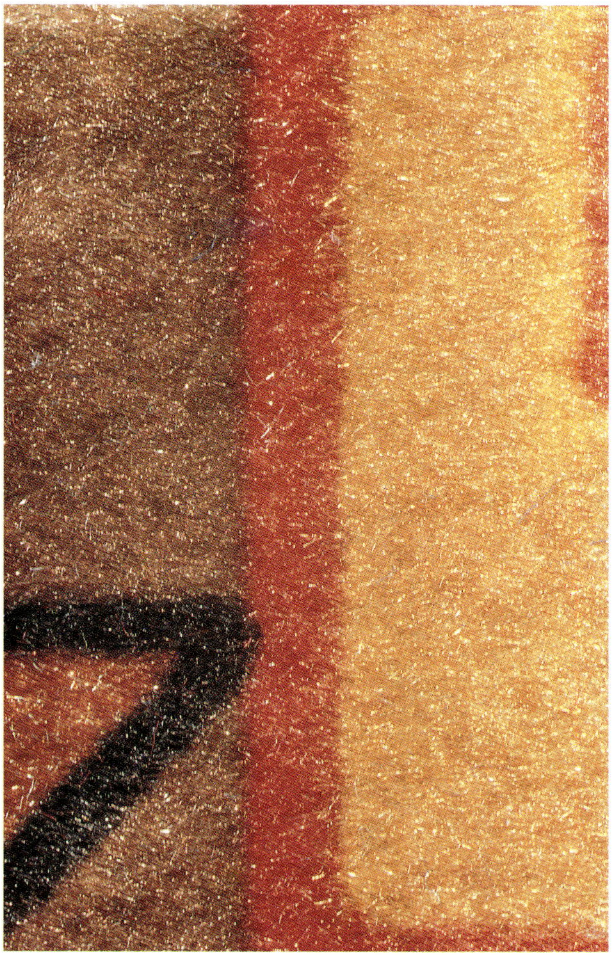

Detail.

Polyester blend print.
Harry D. Spielberg
1972

Opposite page: White, lavender, green, and blue polyester knit with a feather-like motif.
Early 1970s

Polyester multi-kaleidoscope of color, washable print.
Spilke
1975

Synthetic chiffon, multi-color dots and swirling droplets.
Early 1970s

Above: Satin crepe print.
Kanebo
1970

Top right: Detail.

Right: Stretch nylon, white and navy abstract print.
Mid 1970s

Opposite page: Polyester knit in brick red and earth tones.
Mid 1970s

Forest scene in orange, purple, and brown polyester knit.
Late 1970s

Nature inspired, bold stylized tropical leaf print for drapery, bedspreads, and upholstery fabrics.
Design Resources Inc.
1978

Screen print satin.
Kanebo
1970

Opposite page: Polyester abstract print of red and white check with superimposed black shapes with orange insects.
Early 1970s

Opposite page:
Top: Stretch polyester knit, in yellow and pink biomorphic design.
Early 1970s

Bottom: Vibrant floral print, nylon double knit.
Barbet Stretchnit
1970

Top: Silky supply, 100% Enka polyester.
Delden
1975

Bottom: Antron nylon, beautiful velvet print, easy to clean, for upholstery.
DuPont
1978

Top: Polyester knit in pastels.
Mid 1970s

Bottom: Luxurious brushed multi-filament nylon, stitch quilting sleep wear.
Guilford
1975

Vibrant colors, Hawaiian, polyester knit.
Mid 1970s

Bold colors, quilted polyester, with sequins.
Mid 1970s

Floral & Pictorial

Right: Synthetic knit, thin green lines with flower clusters.

Bottom left: Light color, 100% Avlin polyester, solid and floral.
Guilknit
1974

Bottom center: Red, challis hand, fashion knit of Arnel blend.
Wullschleger
1974

Bottom right: Red daisy, 100% polyester double knit, sold in one-stop center by Singer.
1972

Primary colored flowers on yellow-gold background polyester knit. Early 1970s

Trevira, polyester double knit, designed by Soptra's Eric Sorter; a flower garden of color that does not wrinkle. Soptra Fabrics 1970

Top left: Polyester knit, bold vibrant poppies, washable. Mid 1970s

Top right: Pongee - a silk-like polyester print made in Lyon, France. Setila 1978

Bottom left: Polyester knit, camel with small tulips. Late 1970s

Bottom right: Pink nylon floral print for women's robes, made in Holland. Hatéma-Texoprint 1971

Opposite page: Pongee - a silk like polyester print made in Lyon, France. Setila 1978

Opposite page:
Top: Soft pretty Queen Anne print on polyester/ cotton voile.
Wamsutta
1975

Bottom left: Floral print for drapery and upholstery, Kodel polyester, luxurious appearance, light fastness and resists fading in sunlight.
Quaker Fabrics
1977

Bottom right: Ribbed nylon print by Texoprint, one of the largest and finest screen printers in Europe, located in Enschede, The Netherlands.
Texoprint
1970

Swimsuit stretch with Lycra, black with pink & white bold design.
Mid 1970s

100% Fortrel polyester, muted soft water color effect.
Kenmill
1977

Above: Tribal figure with mask motif, polyester cotton blend knit.
Mid 1970s

Top right: Cotton/synthetic blend stretch knit, deep navy and white with red flowers.
Mid 1970s

Right: Cotton/synthetic blend stretch velvet hot pink coat-of-arms motif on gold.
Mid 1970s

Opposite page:
Top: Washable, satin-look synthetic, bold yellow flowers with grey leaves.
Late 1970s

Bottom left: Soft, light weight, heat transfer, printed fabric, double knit polyester.
Coopertex, Inc.
1975

Bottom right: Muted beige, orange, and green polyester bold floral.
Late 1970s

Top: Cotton/synthetic blend knit with jungle motif in bold reds and blues.
Mid 1970s

Bottom: Different area of large pattern.
Mid 1970s

Same pattern in different colorway.

Glossary

Acetate - Manufactured fiber in which the fiber forming substance is cellulose acetate.

Acetylation - Chemical reaction in which cellulose is changed into cellulose acetate.

Air jet - Technique for bulking a filament yarn by subjecting it to jets of air under pressure.

Arnel - Registered trademark of Celanese for its triacetate filament yarn.

Autoclave - A large chamber that can be sealed and filled with steam under pressure.

Avril - Trademark of Avtex Fibers Inc. for their standard, high wet modulus rayon fiber having a round cross section.

Avril III - Trademark of Avtex Fibers Inc. for their natural wet modulus rayon fiber having a multilobal cross section.

Barré - Streaks that develop in a dyed knit fabric made with textured filament yarns.

Bi-component - Man made filament fiber in which two chemical components have been combined.

Bulking - Characteristic result of yarn texturizing.

Chemical cellulose - Carbohydrate of complex molecular structure forming the chief framework of plant cells and walls; the basic raw material of cellulosic fibers.

Coil - Typical spring-like formation of a false twist stretch yarn.

Compacting - Technique for imparting texture to a finished fabric of thermoplastic fibers.

Coning - Winding the textured yarn on a cone, ready for knitting or weaving.

Continuous filament - A long continuous strand of a manufactured fiber as distinguished from all natural fibers (except raw silk), which are of short staple length.

Conventional twist - Original method developed by Herberlein in Switzerland for making stretch yarns by twisting, setting, and untwisting in three stages.

Co-polymer - Man made fiber made with two variants of the same basic chemical polymer.

Crease resistant - Fabrics that have been treated or constructed to improve their resistance to and recovery from wrinkling.

Crimp - General term that describes waviness in fibers, natural or manufactured.

Cross-dyeing - Method of dyeing fabrics containing yarns with different affinities for different dyes.

Cross-linking - Chemical process for modifying the molecular structure of a fiber by forming a permanent bond or link.

Cross-section - Shape of an individual filament when cut at right angles to its axis.

Curl - General term to describe the conformation of a textured filament yarn.

Denier - Unit of weight indicating size of a fiber filament based on thickness (weight) in grams.

Dimensional stability - Tendency of fabric to retain shape and size after wear, washing, and dry cleaning.

Edge crimping - Technique for texturizing a filament yarn by heating it and drawing it over the edge of a sharp blade.

Elongation - Deformation of a fiber due to stretching.

Fabric - Any woven, knitted, plaited, braided, felted, or nonwoven material made of fibers or yarns.

False twist - Yarn is twisted by a rotating spindle and set in a heat box and then it is untwisted. Thus, it is called "false" twist and the twist does not remain.

Feed yarns - Yarns supplied to the throwster by the fiber producer, ready for texturizing.

Fiber - Any substance, natural or man made, composed of thread-like tissue capable of being spun, woven, or knitted.

Filament - An individual strand of continuous length extruded through one hole in the spinneret.

Gauge - Measurement of needles per inch in the knitting machine.

Gear crimping - Yarn texturing method in which the yarn passes through the meshing teeth of two heated gears.

Godet wheel - Wheel over which newly spun man made fiber is wound.

Hand - Term used to describe the touch, handle, or feel of fabrics.

Heat set- To fix or set a yarn in crimped or textured form by use of heat.

Hollow fibers - Man made fibers spun through spinneret with pronounced "C" shaped holes.

Homo-polymer - Man made fiber that consists of one chemical polymer, as distinct from a co-polymer.

Hydrophobic - Property of resisting the absorption of water-common to nylon, polyester, and other synthetics.

Knit-de-knit - Yarn texturing method.

Knitting machines:
Jersey-flat - Basic knitting stitch; the jersey flatbed produces full-fashioned sweaters, as well as, many other types.
Purl-flat - Famous links-links machine, fine for making pullovers and cardigans and the classic golf or tennis sweater.
Jersey-circular - Industry "workhorse," producing underwear, seamless hosiery, piece goods, and other high volume apparel.
Purl-circular - The high volume purl machine, produces cut and sewn sweaters, dress wear and men's socks.
Rib-circular - The classic example is the double knit, the fastest growing knitting machine category for men's wear.
Tricot - Warp knit machine that produces fine vertical lines on the face and crosswise ribs on the back of fabric.
Rib-flat - Complex machine, capable of stitch transfer and it can trim cuffs and collars expertly.
Raschel - Warp knit machine that can stitch or lay in yarn, create open or lacy fabrics, rigid or stretchy fabrics.

Knitting - Process of making fabric by interlocking series of loops of one or more yarns.

Loft - Properties of firmness, resilience, and bulk of a fiber batting, yarn, fabric, or other textile material.

Loom - Weaving machine for producing textiles by interlacing warp and filling yarns in perpendicular fashion.

Loop - The formation given to a yarn by the air jet method of texturizing.

Luster - Shine or sheen of fibers, yarns, or finished fabrics.

Man made cellulosics - Fibers produced by modifying or transforming natural cellulosic polymers.

Modulus - Measure of the stress needed to stretch a fiber to the maximum extent from which it will return completely to its original shape.

Monofilament - Single strand of man made fiber, continuous filament or strand.

Multifilament - Yarns with many fine continuous filaments or strands; multifilament yarns produce especially soft fabrics.

Nonwoven - Material made of fibers in a web or mat held together by a bonding agent or by interlocking fibers as in felt.

Nub yarn - Novelty yarn containing slubs, beads, or lumps intentionally introduced.

Pent-lobal - Man made fiber whose cross section has been modified from circular to five sided.

Permanent press - Pre-cured or post-cured process for producing a garment with a durably set shape and/or designed creases, which requires no ironing after normal laundering for its normal life.

Petroleum based fibers - Fibers produced by modification of polymers derived from petroleum.

Piece dye - Fabric dyed in the piece, after construction.

Producer textured - General term for yarns textured by the fiber producer at the source, rather than by a throwster.

Raschel gauge - Measured as needles per two inches.

Raschel machine - Designed to handle heavier knit spun yarns.

Rayonne - French generic term for viscose rayon staple.

Recovery - Essential property of a stretch yarn.

Resilience - Property of a textile material to recover from a deformed state.

Scroop - Crunching sound or sensation produced by rubbing together fibers having a high degree of inter-fiber, stick-slip friction.

Set - A set yarn is one that has been textured and heat set to remove stretch.

Set yarns - A false twist "set" yarn is really the same stretch yarn that has been given a second heat treatment.

Shrinkage - Treatments to remove most of a fabric's tendency to shrink.

Solution-Dyeing - Process of "dyeing" fiber in the solution stage, before spinning, so that the filament is impregnated with pigment.

Spindle - Rotating device with an eye through which yarn is threaded and then twisted.

Spinneret - Metallic cap or jet with microscopic holes in the surface through which spinning solutions are forced, becoming filaments as they emerge into a coagulating medium.

Stretch - Yarn that has been false twisted and not set.

Stretching - As applied to man made fiber production, the stretching process gives the fiber greater strength after it has been extruded.

Stretch yarns - A filament yarn is twisted 50 or more turns per inch.

Stuffer box - Texturing method by which yarn is stuffed into one end of a heated box and emerges from the other end with a sawtooth crimp.

Synthetic fibers - Fibers manufactured by chemical syntheses.

Tenacity - Measure of yarn strength, i.e., high tenacity denotes high strength.

Tensile strength - Comparative measurement of the strength of fibers, yarns, or fabrics.

Texture - The word derives from the Latin *"texere,"* meaning to weave, but is now applied to crimped yarns and to fabrics with surface interest.

Textured acetate - Acetate filament yarn that has been bulked to produce a more lofty texture.

Thermoplastic - Chemical fiber that becomes malleable when subjected to heat and whose form can be changed permanently by heat.

Throwster - Derived from the old Anglo Saxon word *"thrawan,"* meaning to twist or revolve.

Torque - Stretch factor tendency to curl which is imparted to a yarn by false twisting.

Tow - Large number of continuous filaments drawn and collected in a loose rope-like strand without definite twist, suitable for cutting into staple or flock.

Triacetate - Fiber resists more heat than acetate, generally holds pleats, shape, and texture.

Tricot - French for warp-knit fabric.

Tricot gauge - Measured in needles per one inch.

Tricot machine - Designed to convert smooth, fine denier filament yarns into lingerie and bonding fabric.

Tri-lobal - Man made fiber whose cross section has been modified from circular to three sided.

Twist - Turning of fibers or yarns around their axes, expressed in number of turns per unit of length.

Twist factor - Also referred to as twist multiplier – the ratio of turns per inch to the square root of the yarn count.

Warp crimping - Crimped into a multiple layer of ends arranged in warp formation.

Weaving - Process of manufacturing fabric by interlacing a series of warp yarns with filling yarns at right angles.

Wet-laid - Formed fabrics manufactured by using processes and equipment adapted from paper making techniques.

Wicking - Dispersing or spreading of moisture or liquid through a given area, vertically or horizontally; capillary action in material. The quality of a yarn or fabric that enables it to draw moisture away from the body; thus more comfortable to wear.

Yarn - Assemblage of fibers or filaments, either manufactured or natural, twisted or laid together so as to form a continuous strand that can be used in weaving or knitting, or otherwise made into a textile material.

Selected Bibliography

American Fabrics. New York: Doric Publishing Co., Issues 1970-1979.

American Fabrics. *AF Encyclopedia of Textiles*. 3rd ed. Englewood Cliffs, New Jersey: Prentice-Hall Inc., 1980.

Birrell, V. L. *The Textile Arts: A Handbook of Fabric Structure & Design Processes*. New York: Harper & Row, 1959.

Brown, Patty, and Janett Rice. *Ready-to-Wear Apparel Analysis*. Upper Saddle River, New Jersey: Prentice-Hall, Inc., 1998.

Corbman, Dr. Bernard P. *Textiles: Fiber to Fabric*. New York: McGraw-Hill Inc., 1975.

Davis, Rebecca, and Carol Tuntland. *The Textiles Handbook*. Redondo Beach, California: Plycon Press, 1996.

Demir, Ali, and Hassan Mohamed Behery. *Synthetic Filament Yarn Texturing Technology*. Upper Saddle River, New Jersey: Prentice-Hall Inc., 1997.

Dictionary of Textile Terms. 12th ed. New York: Dan River Inc., 1976.

Garthe, Mary. *Fashion and Color*. Rockport, Massachusetts: Rockport Publishers, 1995.

Gioello, Debbie Ann. *Profiling Fabrics: Properties, Performance & Construction Techniques*. New York: Fairchild Publications, 1981.

Hall, David M. *Practical Fiber Identification*. 2d ed. Auburn, Alabama: Auburn University Press, 1982.

Herald, Jacqueline. *Fashions of a Decade: The 1970s*. New York: Facts on File, 1992.

Hollen, Norma, et al. *Textiles*. 6th ed. New York: MacMillan Publishing Company, 1993.

Houck, Catherine. *The Fashion Encyclopedia*. New York: St. Martin's Press, 1982.

Humphries, Mary. *Fabric Glossary*. Upper Saddle River, New Jersey: Prentice Hall, 1996.

Jerde, Judith. *The Encyclopedia of Textiles*. New York: Facts On File, Inc., 1992.

Joseph, Marjory L. *Essentials of Textiles*. New York: Holt, Rinehart & Winston, 1976.

Kim, Charles J. *Textile Science: An Outline*. Dubuque, Iowa: Kendall/Hunt Publishing Company, 1997.

Lauterburg, Lotti. *Fabric Printing*. New York: Reinhold Publishing Corp., 1959.

O'Hara, Georgina. *The Encyclopaedia of Fashion*. New York: Harry N. Abrams, Inc., Publishers, 1986.

Peacock, John. *The 1970s*. New York: Thames and Hudson, 1997.

Reichman, Charles. *Knitting Dictionary*. New York: National Knitted Outerwear Association, 1966.

Schnurnberger, Lynn. *Let There be Clothes*. New York: Workman Publishing, 1991.

St. James Fashion Encyclopedia: Survey of Style from 1945 to the Present. Detroit, Michigan: Visible Ink Press, 1997.

Tortora, Phyllis B. *Fairchild's Dictionary of Textiles*. Upper Saddle River, New Jersey: Prentice-Hall Inc., 1996.

Tortora, Phyllis B. *Understanding Textiles*. Upper Saddle River, New Jersey: Prentice-Hall Inc., 1996.

Wilson, Kax. *A History of Textiles*. Boulder, Colorado: West View Press, Inc., 1979.

Yeager, Jan. *Textiles for Residential and Commercial Interiors*. New York: Harper & Row, Publishers, 1988.

Index